Recombinant DNA Techniques:
An Introduction

WITHDRAWN

Recombinant DNA Techniques: An Introduction

Raymond L. Rodriguez
University of California
Davis, California

Robert C. Tait
University of California
Davis, California

1983

The Benjamin/Cummings Publishing Company, Inc.
Menlo Park, California · Reading, Massachusetts
London · Amsterdam · Don Mills, Ontario · Sydney

London • Amsterdam • Don Mills, Ontario • Sydney • Tokyo

Library of Congress Cataloging in Publication Data

Rodriguez, R. L. (Raymond L.)
  Recombinant DNA techniques.

  Bibliography: p.
  Includes index.
  1. Recombinant DNA. 2. Genetic engineering--Technique. I. Tait, Robert C. II. Title. III. Title: Recombinant D.N.A. techniques. [DNLM: 1. DNA, Recombinant. 2. Genetic technics. QH 443 R696r]
  QH442.R6 1983   660'.62   83-11864
  ISBN 0-201-10870-4

Copyright © 1983 by The Benjamin/Cummings Publishing Company
Published simultaneously in Canada

All rights reserved. No part of this publication may be reproduced, stored in a retrieval system, or transmitted, in any form or by any means, electronic, mechanical, photocopying, recording, or otherwise, without the prior written permission of the publisher, Benjamin/Cummings Publishing Company, Inc., Menlo Park, California 94025, U.S.A. Printed in the United States of America.

The Benjamin/Cummings Publishing Company, Inc.
2727 Sand Hill Road
Menlo Park, CA 94025

ISBN 0-201-10870-4
BCDEFGHIJ-AL-8987654

*To Wendy, Karen, and
H. W. Boyer*

*He who has imagination without learning
has wings but no feet.*

—Joubert—

Contents

Foreword xiii

Preface xvii

CHAPTER 1 Introduction to Recombinant DNA Technology 1

    Plasmid Cloning Vectors 3

    Single-stranded Bacteriophage Cloning Vectors 9

    Double-stranded Bacteriophage Cloning Vectors 12

    Cloning Strategies 19

    Preface to the Experiments 24

CHAPTER 2 Microbiological Techniques 27

    Antibiotic Resistance 27

    Bacterial Cell Growth 29

Exercise 1 *Aseptic Manipulation and Storage of Bacterial Cultures* 32

CHAPTER 3 DNA Isolation 37

    Prokaryotic Chromosomal DNA 38

    Extrachromosomal DNA 39

    Rapid Isolation of Plasmid DNA 41

    Eukaryotic Chromosomal DNA 42

    Spectrophotometric Assay of DNA Concentration and Purity 42

CONTENTS

Exercise 2 *Isolation and Purification of* E. coli *Chromosomal DNA* 45

Exercise 3 *Determination of the Concentration and Purity of DNA by Ultraviolet Absorption Spectroscopy* 47

Exercise 4 *Large-Scale Plasmid Purification* 48

Exercise 5 *Rapid Isolation of Plasmid DNA (Miniscreen)* 50

CHAPTER 4 Restriction Endonucleases 53
    Host Restriction/Modification 53
    Type II Restriction Endonucleases 55
    Purification of Restriction Endonucleases 58
    Terminating Restriction Endonuclease Reactions 58
    Generating New Restriction Enzyme Cleavage Sites 59

Exercise 6 *Restriction Endonuclease Digestion of Plasmid and* E. coli *Chromosomal DNA* 62

CHAPTER 5 Gel Electrophoresis 67
    Principles of Electrophoresis 67
    Visualizing DNA Molecules by Fluorescence 69
    Interpreting Electrophoretic Patterns 70
    Gel Systems 71
    Gel Artifacts 72
    Tracking Dye 74
    Transfer-Blot/Hybridization Analysis 74

Exercise 7 *Agarose Gel Electrophoresis of Plasmid and Chromosomal DNA* 77

CHAPTER 6 Ligation of DNA 81
    DNA Ligase from *E. coli* and Bacteriophage T4 81
    In Vitro and *In Vivo* Ligation of DNA Fragments 82
    Factors Affecting the Ligation Reaction 83
    Blunt-End Ligation 86

Exercise 8 *Ligation of Restriction Endonuclease Cleaved Chromosomal DNA Fragments and Plasmid Vector DNA* 89

CHAPTER 7 Bacterial Transformation   91

    Conjugation   91

    Transduction   93

    Transformation   93

Exercise 9 *Preparation and Transformation of Competent Cells of* E. coli *K12*   99

CHAPTER 8 Identification and Characterization of Recombinant Transformants   103

    Phenotypic Characterization   103

    Physical Characterization: Plasmid Isolation and Transformation   106

    Physical Characterization: Restriction Enzyme Analysis   106

    Physical Characterization: *In situ* (Colony) Hybridization   109

    Physical Characterization: Immunoassays   109

    Localization of Genes on Recombinant Plasmids   110

    Subcloning   111

    Insertional Inactivation Using Transposon Mutagenesis   113

    Confirming Gene Location   115

Exercise 10 *Phenotypic Characterization of His$^+$ and Ara$^+$ Recombinant Transformants*   119

Exercise 11 *Isolation of Recombinant Plasmid DNA and Retransformation of the Auxotrophic Host*   121

Exercise 12 *Physical Characterization of Recombinant Transformants: Restriction Endonuclease Mapping*   123

Exercise 13 *Determination of Histidinol Dehydrogenase Activity in His$^+$ Transformants*   126

CHAPTER 9 Cloning Regulatory DNA Sequences   129

    Regulatory DNA Sequences   129

    Promoter Structure and Function   130

    RNA Polymerase Binding Studies   131

    Terminators   133

    Promoter and Terminator-Probe Plasmid Cloning Vectors   134

Exercise 14 *Cloning the Promoters for the Arabinose and Histidine Operons of* E. coli   140

Exercise 15 *Using Antibiotic Resistance as an Indirect Measure of Promoter Strength*   146

CONTENTS

Appendices

1 Recipes 149

LB Medium 149
YT Medium 149
M9 Medium 150
Supplementation Levels for Amino Acids 150
Freezing Medium for Bacterial Strains 151
Yeast Minimal Medium 151
YEPD (Yeast) Medium 153
YNBD (Yeast Minimal) Medium 153
10X Tris-Borate Buffer 153
Phenol Equilibration Buffer 153
A50 Buffer/Dowex Buffer 153
TEN (DNA) Buffer 154
10X SSC (Standard Saline-Citrate) 154
Dowex Preparation 154

2 Protocols 155

Plasmid DNA Isolation Using the CsCl Dye-Buoyant Density Gradient Method 155
Isolation of High Molecular Weight Plasmid DNA 159
Miniscreen for Large Plasmids 160
Isolation of *Bacillus subtilis* Chromosomal DNA 162
Rapid Isolation of Plasmid DNA (Miniscreen) for *Bacillus subtilis* 164
Purification of Phage Lambda DNA 165
Isolating High Molecular Weight DNA from Yeast 167
Isolation of Yeast Plasmid DNA 170
Rapid Method for Isolating Plasmid DNA from Yeast 171
Drosophila DNA Isolation 172
Extraction of Restriction Fragments from Low-melt Agarose Gels 173
Preparative Agarose Gel Electrophoresis 175
Electroelution of DNA Fragments from Polyacrylamide Gels 178
Transformation of *E. coli* χ1776 179
M13 Transformation 180
Bacteriophage Lambda *In Vitro* Packaging 181
Agrobacterium tumefaciens Transformation 183
Transformation of *Bacillus subtilis* with Plasmid DNA 184

Transformation and Storage of Competent Yeast Cells   186
Assay for Chloramphenicol Acetyltransferase (CAT) Activity   187
Maxicell Purification and Labeling of Proteins   192
Minicell Purification and Labeling of Proteins   194
SDS-polyacrylamide Denaturing Protein Gels   197
Preparation of Dialysis Tubing and Minidialysis Chamber   200

3 Restriction Maps and Nucleotide Sequences   203

Restriction Map of pBR322   204
Nucleotide Sequence of pBR322   205
Restriction Map of pBR327   207
Nucleotide Sequence of pBR327   208
Restriction Map of pBR328   210
Restriction Map of pBR329   211
Nucleotide Sequence of pBR329   212
Restriction Map of M13mp7   214
Nucleotide Sequences of M13mp7, mp8, mp9, mp10, and mp11   215
Nucleotide Sequences of pUC7, pUC8, and pUC9   216
Restriction Map of Lambda   217
Restriction Map of pUB110   218
Restriction Map of pBD9   219

4 Tables   221

DNA Polyacrylamide Slab-gel Reagent Volumes   221
Antibiotic Resistant Transposable Elements   222
Nucleic Acid Modifying Enzymes   224
10X Enzyme Reaction Buffer   226
Restriction Endonucleases   227
Supplies and Equipment   230

Index   233

Foreword

Genetic engineering is a concerted application of various broad disciplines and specific techniques to solving basic and practical problems concerning living organisms. The practical problems include improvement of the characteristics of economically important organisms and their products with the purpose of benefiting humankind and its environment. The earliest known examples are the breeding and adaptation of wild plants and animals to the service of another species. Such selective breeding was the basis for development of agriculture and animal husbandry, which probably originated over 17,000 years ago in the valley of the Nile near Aswan. Here, the surrounding desert prohibited unrestricted two-dimensional migration in the search for food, which forced the inhabitants to use their inborn ingenuity to cope with periods of famine. Thus humans changed from a hunter and gatherer to a farmer.

The recombinant DNA technique is one of the most recent developments in the field of genetic engineering, originally motivated not by famine but by intellectual curiosity and the constant quest for knowledge. Recombinant DNA techniques permit isolation and propagation of individual genes, the study of their structure and function, their transfer to various species, and efficient expression of their products. Such versatility offers enormous benefits for the basic and applied sciences, including better understanding of the biology of the lower and higher organisms, advances in medicine and food production, revolutionary improvements in various industrial processes, and cleaning up the environment.

In view of these immense benefits to humankind, it was both tragic and pathetic that early developments in this field were greatly hampered by apprehensions about imaginary dangers of the recombinant DNA technique. These unfounded apprehensions led to unnecessary and onerous regulations, some of which acquired a life of their own, despite the fact that the "risks" of gene combinations created by the recombinant DNA techniques are presently assessed as insignificant from the practical point of view. Both the inadvertent creation of harmful gene combinations and their survival in the natural environment are now considered unlikely. Actually, a major effort in the recombinant DNA technology is to preserve the poorly adapted laboratory-created gene combinations, survival of which depends on special care under specially designed artificial conditions.

Any regulations of scientific endeavor should be based on the careful evaluation of the risks-versus-benefits of both the actual research activity and the regulatory countermeasures. This was not done before the introduction of the recombinant DNA regulations. There is no logical justification of any special regulation for this technique in view of (1) the absence of any known risks of the technique, (2) the insignificance of the hypothetical risks, (3) the enormous benefits of the techniques, (4) the known and unforeseen counterproductive aspects of the regulations, including the violation of the freedom of inquiry, and (5) the insignificant benefits of the bureaucratic regulations.

Transfer of individual genes within or across species boundaries occurs in nature. In the laboratory, special techniques were developed to use natural vectors, e.g., lysogenic bacteriophages, mainly lambda, and transposing elements, to isolate, propagate, and study individual genes such as *lac, gal, bio,* and others. The recombinant DNA technique, employing cutting and splicing enzymes, was the natural extension of these earlier techniques. There was much private discussion in the late 1960s about applying DNA ligase in making new gene combinations, while the members of H. G. Khorana's laboratory were systematically constructing an artificial gene by ligating together synthetic oligonucleotides of predetermined sequences. In 1968–69 I had planned to rearrange phage genes employing shearing and the enzyme T4 ligase and, moreover, to clone foreign genes, including human DNA, in the central portion of the phage λ DNA in cooperation with Dr. V. Sgaramella, who had just discovered blunt-end ligation by T4 DNA ligase in H. G. Khorana's laboratory. However, the moving of Khorana's group to M.I.T. precluded this cooperative venture at that time.

The recombinant DNA techniques, as presently known, were developed in the early 1970s and were based on the use of vectors, mainly plasmids or viral DNAs, in analogy to the natural cloning that employed transducing bacteriophage lambda vectors. Discovery and availability of restriction enzymes for cutting DNA at specific sites, in addition to the ligase-mediated splicing, and the development of useful vectors have placed the method within relatively easy reach of any molecular geneticist. The senior author of this text has made many major contributions to the early development, improvement, and simplification of the technique, including the construction of the famous and widely used plasmid vector pBR322. Seldom in science has the introduction of a new technique had such an overwhelming and immediate impact on science in general and biology in particular. The human spirit of discovery and the desire to help humankind have triumphed over the illogical doomsday scenarios and bureaucratic regulations. The authors should be congratulated on the excellence of their major scientific contributions to the field and this very helpful and comprehensive manual.

FOREWORD

Waclaw Szybalski
Editor-in-Chief, *Gene*
McArdle Laboratory
  for Cancer Research
University of Wisconsin
Madison WI 53706 (U.S.A.)

Preface

In the past decade, we have witnessed a revolution in the fields of molecular genetics and biochemistry. This revolution has not been the result of a single new development in instrumentation or a theoretical breakthrough, but rather due to the application of a variety of techniques collectively referred to as "Recombinant DNA Technology." This technology involves the *in vitro* modification and recombination of genetic material from different organisms to create new gene combinations.

The purpose of recombinant DNA research is to study gene structure and function, and, in many instances, to produce gene products that are otherwise difficult to obtain in pure or large amounts. Due to the broad spectrum of biological problems amenable to the recombinant DNA approach, the scientific literature is filled with exciting and significant reports describing new aspects of gene regulation, the organization of the eukaryotic genome, and the synthesis of mammalian proteins in bacteria. As a consequence, there has been a growing demand, among scientists and students alike, for the knowledge and training needed to apply this experimental approach to their own research projects.

In the fall of 1979, my colleagues and I at the University of California, Davis, offered a ten-week course entitled "Advanced Molecular Genetics Laboratory." The participants in this course included both graduate students and undergraduates with diverse backgrounds in the biological sciences. The purpose of this course was to acquaint these students with the concepts and techniques involved in recombinant DNA research, and their application to the genetic analysis of the arabinose and histidine operons of *E. coli*. These particular operons were chosen because of the wealth of information already available on their regulation and genetic organization. Furthermore, they represent interesting examples of positive and negative mechanisms for regulating gene expression. In response to the heavy demand for copies of laboratory handouts and protocols, we decided to make this material available in published form.

This manual is not intended to be a "cookbook" describing all recombinant DNA techniques currently available, but rather to provide the basic laboratory experience to allow the student to progress to more

PREFACE

advanced experiments. The laboratory exercises have been arranged so as to build on information covered in the preceding exercise. However, each exercise is presented in the context of a chapter, which discusses the theoretical and practical aspects of the experiment. Therefore, each chapter represents a module that can be used to perform the exercises, independently from the other chapters. The exercises are designed to instruct the students in the techniques of transformation, ligation, use of restriction enzymes, and the purification and analysis of DNA. Since our course has attracted students with varied backgrounds and a wide range of interests, the introductory sections to each chapter have been written in such a way as to be intelligible to those with a general knowledge of either microbiology, genetics, or biochemistry. The manual includes extensive appendices where recipes, reagents, buffers, and additional protocols can be found. In short, these features make the manual not only useful for laboratory instruction, but also as a reference book for individual research interests. Upon completion of these exercises, the student or researcher should have the necessary skills and understanding needed to proceed to more advanced and specialized methodologies.

All of the experimental procedures described in this manual have been performed by the authors and have been found to yield clear and reproducible results. Protocols have been carefully written so that the exercises can be performed in a minimum amount of time by the inexperienced experimentalist. Furthermore, the experiments utilize only a small number of commercial restriction enzymes and common laboratory equipment. Since these experiments involve the formation of recombinant DNA molecules from organisms that exchange genetic material naturally, they are exempt from the National Institutes of Health Guidelines on Recombinant DNA Research. However, in accordance with the Guidelines, these experiments should be performed in a P1 physical containment facility using the appropriate laboratory practices.

Due to the broad scope of the recombinant DNA field, it is impossible for us to cover all aspects of molecular cloning technology. For this reason we have focused our attention on plasmid vector systems and bacterial genes. We believe that these experiments are representative of many of the molecular cloning experiments currently found in scientific literature. We hope that this manual will fill the void that presently exists between the scientific literature and existing laboratory manuals.

Finally, we would like to acknowledge the many students and colleagues who have participated directly or indirectly in the development of this text. We would like to thank Robert W. West, Jr., Timothy J. Close, David S. Goldfarb, Laurie A. Quinn, and Byron E. Froman for their help in designing and performing many of the experimental techniques presented here. A special debt of thanks goes to Crystal DiModica and Candy Miller for their help in manuscript preparation, and to Tina L. Shepherd for her artistic and photographic services.

Raymond L. Rodriguez
Robert C. Tait
Department of Genetics,
University of California, Davis

CHAPTER 1

Introduction to Recombinant DNA Technology

Genetic engineering, also known as recombinant DNA research or cloning, is not a discipline *per se* but rather a very powerful investigative tool that has helped revolutionize our concept of the gene. Genetic engineering can be defined as any deliberate manipulation of genes within or between species for the purpose of genetic analysis or strain improvement. A definition this broad must include the plant and animal breeders, classical geneticists, and molecular biologists. Indeed, one of the first usages of the term *genetic engineering* was by Hansche et al., who reported in the October 1971 issue of *Diamond Walnut News* that "Our accumulated knowledge of the mechanisms of heredity and the demands for efficiency have turned varietal improvement into a problem of biological engineering and the plant breeder into a genetic engineer." Recombinant DNA technology is a more specialized category of genetic engineering. This area of study involves the *in vitro* rearrangement of genetic material by enzymatic manipulation. These *in vitro* rearrangements produce recombinant molecules composed of at least two different pieces of DNA. Generally speaking, the resulting recombinant molecule would never be formed under natural circumstances. Regardless of how clever or socially relevant these particular gene arrangements may appear, they are of little significance until they are introduced into a biological host where they can be maintained and propagated indefinitely.

As can be seen in Fig. 1.1, a typical cloning experiment requires: (1) DNA of interest, sometimes called foreign, passenger, or target DNA; (2) a cloning vector; (3) restriction endonucleases; (4) DNA ligase; and (5) a prokaryotic or eukaryotic cell to serve as the biological host. Once the vector and foreign DNA have been isolated, they are treated with the same restriction endonuclease to produce site-specific scissions in the DNA. Many restriction enzymes produce double-stranded cuts separated by four base pairs. Therefore, the resulting DNA fragments will possess overlapping single-stranded ends. Since the ends of the vector and the

CHAPTER 1
INTRODUCTION TO
RECOMBINANT DNA TECHNOLOGY

TYPICAL CLONING EXPERIMENT

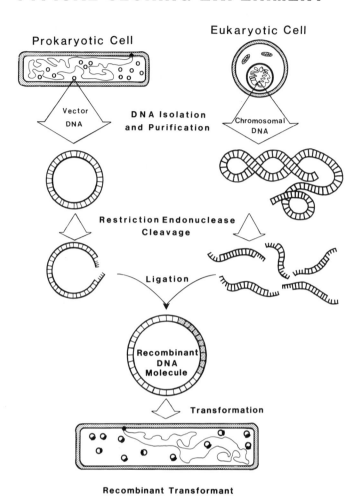

Figure 1.1
A typical recombinant DNA experiment depicting the cloning of eukaryotic genomic DNA fragments in *E. coli*.

foreign DNA are complementary (when generated by the same restriction enzyme), they can anneal, thus producing a recombinant molecule. These new molecules are made permanent only when all gaps are sealed by DNA ligase. Once this is accomplished, the recombinant molecules can then be physically inserted into the appropriate host by transformation or microinjection. The recombinant molecules must then be isolated and characterized. It is through the characterization process, which will be described in detail in subsequent chapters, that most of the recent advances in our knowledge of gene structure have been obtained. The cloning process itself merely provides a means by which genes can be quickly identified and isolated.

It has been suggested that many DNA arrangements that have been constructed *in vitro* can also be constructed using more traditional genetic approaches. If this is true, what is the advantage in using recombinant

DNA technology? The answer lies in the frequency of occurrence of the desired recombinant molecule. For example, it may be possible to obtain a bacterial plasmid that contains the human insulin gene by transforming bacteria with human DNA and allowing illegitimate recombination to construct the desired molecule. However, the frequency of such an event may be so low as to make the isolation and identification of the correct recombinant difficult if not impossible. With *in vitro* manipulations of purified or partially purified genes, the chances of constructing the desired molecule can be increased by several orders of magnitude, making the isolation and identification of the desired product technically feasible. Because of the speed and specificity of the *in vitro* approach, DNA molecules that may occur naturally at an extremely low frequency, if at all, can be constructed easily and efficiently. It is the accelerated nature of genetic analysis by cloning that researchers find most appealing.

Plasmid Cloning Vectors

As shown in Fig. 1.1, the molecular cloning vector plays an important role in recombinant DNA experiments. Since most DNA fragments are incapable of self-replication, an autonomously replicating segment of DNA must be used to accomplish replication of the desired DNA fragment. Most cloning vectors have been derived from extrachromosomal replicons such as bacteriophage, viruses, and plasmids. Plasmids are small double-stranded circular DNA molecules that are capable of replicating in bacteria independently of the bacterial chromosome (fig. 1.2). Because they contain genes for a variety of functions, plasmids have been called the "genetic spare change" of bacteria.

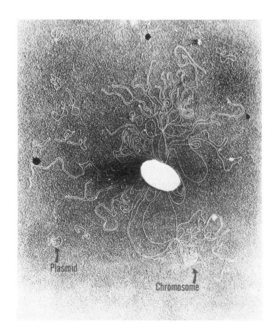

Figure 1.2
An electron micrograph showing chromosomal and plasmid DNA molecules extruded from a single bacterial cell. (Courtesy of H. Potter and D. Dressler, 1980.)

CHAPTER 1
INTRODUCTION TO RECOMBINANT DNA TECHNOLOGY

The ancestry of most plasmid cloning vectors in use today can be traced back to a small number of R-factors (plasmids that encode resistance to various antibiotics) and antibiotic resistant transposons. The process of constructing a plasmid vector usually involves a number of *in vivo* and *in vitro* manipulations, which eventually produce a plasmid with all the requisite features of an efficient plasmid cloning vector. While some of these features may not have been anticipated in the early stages of the construction, most researchers adhere to the following criteria to guide the design of the vector.

Criteria for Plasmid Vector Design

1. A plasmid vector should be as small as possible with little or no extraneous genetic information. Since the efficiency of transformation of many cell types decreases as the size of the plasmid increases above 15 kilobases (kb), the vector should contribute as little as possible to the overall size of the recombinant molecule.
2. Vectors should be well characterized with respect to gene location, restriction enzyme cleavage sites, and nucleotide sequence (if possible).
3. The vector should be easily propagated in the desired host so that large quantities of vector and recombinant DNA molecules can be obtained.
4. The vector should have a selectable marker (gene) that allows cells transformed with the vector to be distinguished from nontransformed cells.
5. An ideal vector should have an additional genetic marker that can be inactivated by the insertion of a foreign DNA fragment (insertional inactivation). The inactivated gene will allow cells harboring recombinant molecules to be distinguished from nonrecombinant molecules on the basis of altered phenotype (fig. 1.3).
6. Finally, the vector should possess the maximum number of unique restriction endonuclease cleavage sites located in one or the other genetic marker. This provides maximum flexibility in terms of cloning different kinds of restriction fragments.

An example of how a plasmid vector is constructed is shown in Fig. 1.4a and b. This figure illustrates the various steps involved in the construction of the plasmid pBR322 and its plasmid derivatives. The salient points in the construction pathway involve the acquisition of (1) a "relaxed mode" of plasmid DNA replication, and (2) the ampicillin and tetracycline resistance genes to serve interchangeably as genetic markers for selection and insertional inactivation.

For the gram-negative bacteria, plasmid replication generally falls into two basic modes: stringent and relaxed. Stringent replicating plasmids, such as pSC101, require protein synthesis and the activity of DNA polymerase III. These plasmids are usually present in a few copies (1–5) per cell. On the other hand, relaxed replicating plasmids, such as ColE1, require the activity of DNA polymerase I and can replicate in the absence of protein synthesis. These plasmids are usually present in 30 to 50 copies per cell, and upon inhibition of protein synthesis, they will continue to replicate while chromosomal replication ceases. This phenomenon, known as amplification, results in approximately a 100-fold increase in cellular plasmid DNA.

Starting with a clinical isolate harboring the colicin-producing plasmid pMB1, the relaxed origin of DNA replication was isolated in the form of pMB8. The tetracycline resistance (Tc^r) gene from pSC101 and the ampicillin resistance (Ap^r) gene from the transposon Tn3 (carried on the plasmid pRSF2124) were incorporated into the plasmid pBR312. All subsequent manipulations were designed to maximize the number of unique restriction enzyme cleavage sites and to minimize plasmid size. For example, the extra *Bam*HI site near the Ap^r gene in pBR312 was removed by *Eco*RI* digestion (see Chapter 4). Likewise, the two additional *Pst*I cleavage sites in pBR313 were removed by a more complicated maneuver to give rise to pBR322. As can be seen in Fig. 1.4a, pBR322 contains nine unique restriction enzyme cleavage sites, five of which reside in either the Tc^r or Ap^r genes. Since the enzyme *Hinc*II also cleaves the *Sal*I site, it is not included among the unique cleavage sites (see Appendix 3 for a restriction map of pBR322).

With the exception of pBR327 (fig. 1.4b), which lacks 1089 base pairs (bp) of nonessential DNA present in pBR322, subsequent plasmid constructions were designed to introduce a unique *Eco*RI cleavage site within a plasmid gene. In this way, *Eco*RI restriction fragments could be cloned and identified by the insertional inactivation of a plasmid encoded gene.

Insertional Inactivation of the Tetracycline Resistance Gene

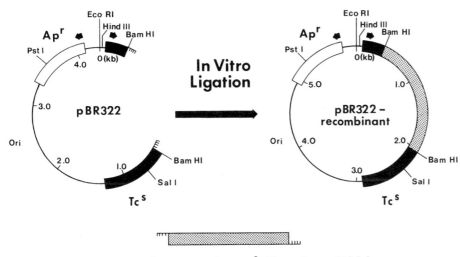

Figure 1.3
Diagrammatic representation of insertional inactivation of the tetracycline resistance (Tc^r) gene of pBR322. Ap^r = ampicillin resistance, Tc^s = tetracycline sensitivity.

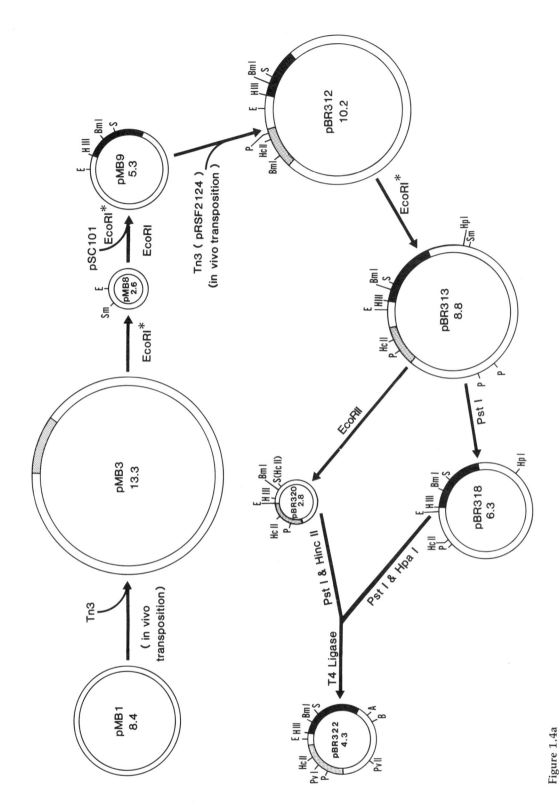

Figure 1.4a
Schematic representation showing the steps in the construction of pBR322 from pMB1, pSC101 and pRSF2124. A, *AvaI*; B, *BalI*; BmI, *BamHI*; E, *EcoRI*; HIII, *HindIII*; HcII, *HincII*; HpI, *HpaI*; P, *PstI*; PvI, *PvuI*; PvII, *PvuII*; S, *SalI*; Sm, *SmaI*. (Adapted from 25, F. Bolivar, "Molecular Cloning Vectors," 1979, Pergamon Press, Ltd.)

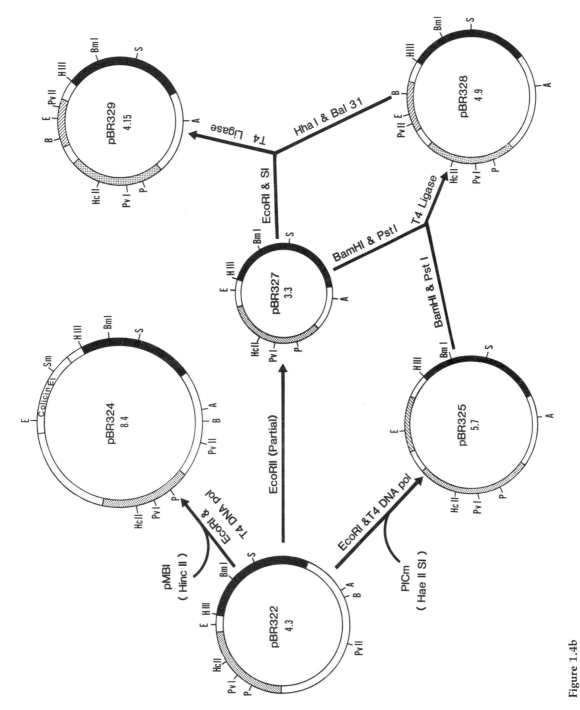

Figure 1.4b
Schematic representation showing the steps in the construction of pBR329 from pBR322, pBR325, and pBR328. Solid bar = Tc^r gene; stippled bar = Ap^r gene; hatched bar = Cm^r gene. (Adapted from *Life Sciences* vol. 25, F. Bolivar, "Molecular Cloning Vectors," 1979, Pergamon Press, Ltd.)

CHAPTER 1
INTRODUCTION TO RECOMBINANT DNA TECHNOLOGY

This was accomplished by introducing into pBR322 the colicin E1-production gene from pMB1 (pBR324) and the chloramphenicol resistance (Cm^r) gene from bacteriophage P1 (pBR325). Due to the convenience and reliability of the selection for Cm^r cell growth, this gene has received more attention and study. For example, by exchanging *PstI/BamHI* restriction fragments between pBR325 and pBR327, the Cm^r gene was transferred to a smaller plasmid (pBR328). However, it was soon discovered that in the process of constructing pBR325, the carboxy-terminal end of the Tc^r gene had been duplicated at the end of the Cm^r gene. Since a duplication of this sort may lead to plasmid instability and loss of cloned fragments, it was believed that this duplication posed a potential obstacle to the use of pBR325 as a cloning vector. This problem was resolved when the duplicated region at the end of the Cm^r gene was enzymatically removed with the nuclease *Bal*31 to generate pBR329 (see Appendix 4 for a table of DNA modifying enzymes).

It should be noted that in the course of removing the duplicated portion of the Tc^r gene in pBR328, the orientation of the Cm^r gene was reversed in pBR329. It was later found that the *Bal*31 treatment had also removed the transcriptional start-point (i.e. promoter) for the Cm^r gene, and that this orientation placed the structural portion of the gene under the control of the anti or α-Tc promoter. Although the function of this promoter is unknown, it is known to overlap and converge with the promoter for the Tc^r gene (fig. 1.5). In the case of pBR329, the α-Tc promoter effectively transcribes the Cm^r gene *in vivo*, albeit to a lesser extent than the Cm^r promoter.

In addition to the pBR322-series plasmids, there are a number of other vectors that can be used for a variety of cloning purposes. Some of these vectors are designed to extend the cloning technology to organisms other than *E. coli*. For example, there are cloning vectors that function in a variety of bacteria as well as in eukaryotes, such as yeast and mam-

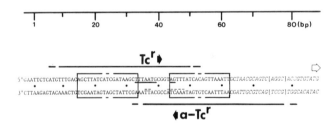

Figure 1.5
Diagram of Tc^r promoter region. The region shown includes the sequence extending from position 1 to position 88 of the wild-type pBR322 nucleotide sequence. The promoter designations (Tc^r and α-TC^r) are placed above or below the putative firm-binding sites for RNA polymerase, with the directions of transcription denoted. The Pribnow box (5'-TTTAATG-3') and initiation site (5'-AG-3') for the Tc^r promoter are underlined. For the α-Tc^r promoter, the possible Pribnow sequence (5'-TAAACTA-3') and initiation site (5'-TT-3') are indicated above the sequence with broken lines. Two regions of dyad symmetry (19 bp, with 3-bp hyphenation), representing potential sites of binding by the Tc^r repressor are indicated by open boxes. The open arrow indicates the starting point for translation of the Tc^r gene product. (Modified from West and Rodriguez, Gene, 20:291, 1982.)

malian tissue culture cells. Still other vectors are designed to clone specific segments of DNA, such as regulatory DNA sequences. A summary of the different types of molecular cloning vectors will be provided later in this chapter.

As mentioned, the detection of recombinant DNA molecules can be achieved using cloning systems that involve the insertional inactivation of vector-encoded genetic markers. Messing and co-workers have developed two such systems that employ the *E. coli lacZ* gene product, β-galactosidase (β-gal), and a chromogenic compound (X-gal) that turns blue when cleaved by β-gal. One system utilizes the male specific single-stranded (ss) DNA bacteriophage M13, while the other is based on the pUC derivatives of the plasmid pBR322 (see Appendix 3). With the M13 system, recombinant bacteriophage can be distinguished from nonrecombinants by plaque color. Colorless or white plaques are indicative of recombinant phage, while blue plaques indicate a nonrecombinant phage. With the pUC plasmids, recombinant and nonrecombinant plasmid molecules can be identified as white and blue colonies, respectively.

The M13 phage enter the male bacteria through proteinaceous filaments called "sex pili." Once in the cell, the plus (+) strand of the virus is used to synthesize a complementary minus (−) strand, which gives rise to a double-stranded copy of M13 known as the parental replicative form (RF). The parental RF serves as a template from which 100 to 200 copies of progeny RF molecules are synthesized. While progeny RF are being synthesized, the (−) strand, which also serves as the viral sense strand, is being transcribed and the RNA translated into viral gene products. One of these gene products (gene V) interferes with M13 RF synthesis in a way that leads to the accumulation of (+) strand DNA. The (+) strand/gene V protein complex makes its way into the periplasmic space of the cell where it is assembled with viral coat proteins into a mature phage particle. The phage particles are then released from the cell without disruption of the cell wall or harm to the bacteria. However, the infectious process does retard cell growth, and as M13 phage particles continue to infect surrounding male bacteria, a "plaque" of slow-growing cells is formed. It should be mentioned that mature phage particles are not the only means for establishing phage infection. The infectious process can also be initiated by transforming male bacteria with single-stranded or RF M13 DNA. No packaging of viral DNA is required with this phage vector system.

So how do the nonrecombinant M13 phage vectors (mp2–11 series) produce blue plaques? This is accomplished by designing M13 and its host in such a way as to take advantage of the phenomenon known as α-complementation. The *E. coli* β-gal protein is composed of 1,021 amino acids, and is active as a tetramer. Certain deletions in the *lacZ* gene near the amino-terminal end produce an inactive enzyme that can be complemented by deletions in the carboxy-terminal region of the gene. The latter class of deletions produces a truncated peptide of β-gal called the α-donor, while the amino-terminal deletions produce what is called the α-acceptor. Alpha complementation is achieved when the α-donor and α-acceptor aggregate *in vivo* or *in vitro* to restore β-gal activity.

CHAPTER 1

INTRODUCTION TO RECOMBINANT DNA TECHNOLOGY

Single-stranded Bacteriophage Cloning Vectors

Due to the presence of a noncoding or intergenic region between gene II and gene IV (fig. 1.6), it was possible to introduce a portion of the *lac* operon into M13 without affecting phage viability. As shown in Fig. 1.6, the insert contains the *lac* promoter, operator, the amino-terminal end of β-gal (i.e., the α-donor peptide). Within the coding sequence for the α-donor peptide, a number of unique restriction enzyme cleavage sites or multiple cloning sites (MCS) were introduced. The presence of these sites does not affect the α-complementing ability of the α-peptide.

When this M13 derivative is used to infect a strain containing a *lacZ* gene from which the α-region of β-galactosidase has been deleted, the inactive α-fragment produced by M13 can interact with the inactive β-galactosidase produced by the cell to produce an octomeric aggregate that possesses β-galactosidase activity (fig. 1.7). When plated on medium containing the compound 5-bromo-4-chloro-3-indolyl-β-D-galactopyranoside (X-gal) and the inducer isopropyl-β-D-thiogalactopyranoside (IPTG), cells that contain galactosidase activity will turn blue, while those that do not will remain white. When the M13 vector is present in the strain JM101 ($r_k^+ m_k^+$) or JM103 ($r_k^- m_k^+$), α-complementation occurs and blue plaques are generated by the infected cells. When a fragment of DNA is inserted into one of the unique cloning sites in the M13 vector, production of the α-peptide is prevented and no α-complementation is observed (fig. 1.8). Plaques generated by recombinant phage generally remain clear or white in appearance in the presence of X-gal and IPTG.

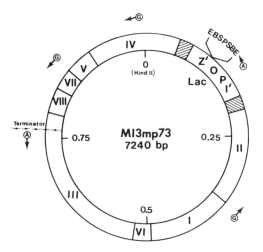

Figure 1.6
Genetic and partial restriction maps of the M13 cloning vector mp73. Roman numerals I through VII denote phage-encoded genes while the numbers 0 to 0.75 indicate the percent of the circular genome. Capital As and Gs encircled with arrows represent the first nucleotide at the 5' end of the different phage transcripts (i.e., phage promoters). All counterclockwise transcription on mp73 terminates after gene VIII. EBSPSBE represents the multiple cloning sites in the α-region of the gene for β-galactosidase (Z'). E, B, S, and P stand for *Eco*RI, *Bam*HI, *Sal*I and *Pst*I respectively. Parentheses around *Hind*II indicates that this site has been removed by mutation. P and O refer to the promoter and operator for the *lac* operon while I' represents a remnant of the *lac* repressor gene. The hatched regions indicate the intergenic region on M13.

Figure 1.7
A representation of the genetic composition of JM101. Filaments on the surface of the cell represent sex pili encoded on the F-factor. Δ = deletion.

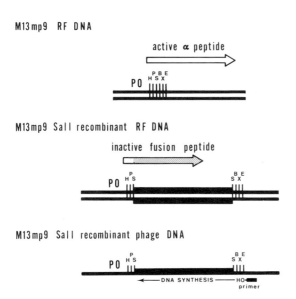

Figure 1.8
Mechanism of insertional inactivation of the α-complementing peptide in M13mp9. H, P, S, B, X, and E refer to *Hin*dIII, *Pst*I, *Sal*I, *Bam*HI, *Xma*I, and *Eco*RI, respectively. The thick lines represent a target DNA fragment inserted at the *Sal*I site. A 17 base single-strand primer is used to initiate DNA synthesis in the dideoxy-DNA sequencing reaction.

CHAPTER 1

INTRODUCTION TO RECOMBINANT DNA TECHNOLOGY

The pUC plasmid vectors are deletion derivatives of pBR322 that contain the Ap^r gene and the same portion of the *lac* operon (with multiple cloning sites) described previously. As with the M13mp vectors, the ability of the α-donor peptide to promote α-complementation is abolished by the insertion of a DNA fragment into one of the MCS. Therefore, when a mixture of recombinant and nonrecombinant pUC plasmid molecules are used to transform JM101 and the cells plated on the appropriate medium, Ap^r transformants harboring the two types of plasmids can be distinguished as white and blue colonies.

Finally, due to the ease and convenience of the dideoxy-method for rapid DNA sequence analysis, the M13mp cloning system has become increasingly popular. This sequencing method, developed by F. Sanger and co-workers, requires single-stranded DNA to serve as a template in a series of DNA-chain-termination reactions. Since only the (+) strand of M13 is packaged and extruded from the cell, mature phage particles provide a convenient source of single-stranded DNA. Because M13 RF and plasmid DNA can be isolated by similar means, a source of double-stranded DNA (for manipulations involving restriction enzymes) is also available.

Double-stranded Bacteriophage Cloning Vectors

Bacteriophage lambda (λ) is one of the most studied and best understood extrachromosomal elements in bacterial genetics. Over the past twenty years, λ has served as a model system for the study of bacteriophage morphogenesis, DNA replication, and gene regulation. Therefore, it is not surprising that it would also be called upon to serve as one of the first molecular cloning vectors.

Bacteriophage λ is a temperate as opposed to virulent phage because its life cycle can be divided into two distinct phases: lytic and lysogenic. The lytic phase of the cycle begins with the injection of λ DNA into the bacterial host cell. Due to the presence of a 12 base, cohesive-end site (*cos*) at each end of the linear molecule, the phage DNA assumes a circular conformation and proceeds to replicate. During the infectious process, the λ DNA molecules monopolize the biosynthetic machinery of the host cell to produce progeny phage consisting of linear double-stranded DNA encapsulated in a protein coat. In order for this protein/DNA complex to mature into viable phage, the coat proteins must encapsulate at least 75% (36.8 kb), but no more than 105% (51.5 kb), of the normal phage genome. This is accomplished by the action of site-specific nucleases located at the base of the phage head that cleave at the *cos* sites. As will be discussed later, the requirement for a "head-full of DNA" is an essential feature of the λ vector system. Upon completion of the lytic process, the host cell wall disrupts, releasing about 100 progeny phage particles.

In the lysogenic mode, the infecting bacteriophage DNA inserts itself in a harmless fashion into the host chromosome. Here, the λ genome is maintained in a passive state by chromosomal DNA replication. The decision to lyse or lysogenize the bacterial host is a complex one, involving a delicate balance in the synthesis of three phage encoded genes: *cI*, *cII* and *cro*.

Another feature of λ that makes it useful as a cloning vector is the distribution of genes along the phage genome. As shown in Fig. 1.9, the genes required for the lysogenic phase of the life cycle (*att, int, xis,* and *red*) are clustered in the middle of the genome spanning two *Eco*RI fragments (fragments B and C, fig. 1.10). The genes essential for lytic growth are located on the left and right sides of the genome and do not fall within *Eco*RI fragments B and C. Since fragments B (5.5 kb or 11.2%) and C (5.9 kb or 12.0%) comprise only 23.3% of the λ genome, phage strains lacking these fragments do not violate the head-full rule and will be capable of lytic growth. By selecting for nonlethal mutations on the right side of the λ genome, a phage mutant (λgt) was obtained that lacked the two rightward *Eco*RI sites. This allows the region occupied by fragments B and C to be replaced with "stuffer" DNA ranging in size from 0 to 14 kb (28%). For example, *Eco*RI fragments from any source could be ligated *in vitro* to the left and right arms of λgt and the products used to transfect *E. coli* cells. The resulting plaques represent a mixed population of recombinant (carrying 0–14 kb inserts) and nonrecombinant phage (λgt-0). To eliminate the nonrecombinant portion of the phage population, the *nin*5 deletion was incorporated into the rightward

CHAPTER 1

INTRODUCTION TO RECOMBINANT DNA TECHNOLOGY

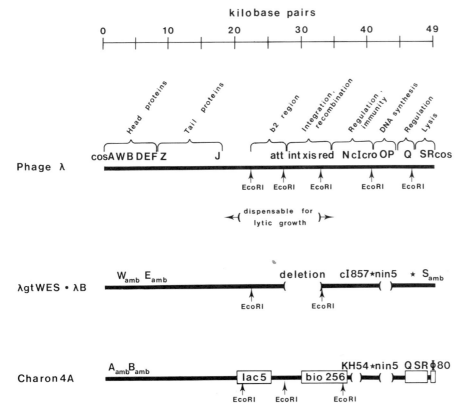

Figure 1.9
Genetic and physical map of the bacteriophage λ. Two λ cloning vectors, λgtWES·λB and Charon 4A, are shown below. Stars indicate the loss of *Eco*RI sites by mutation while amb indicates the presence of an amber (UAG) nonsense mutation in various λ genes. The designation cI857 denotes the temperature sensitive λ repressor gene and *lac* 5, bio 256 are segments of the *lac* operon and biotin gene from the *E. coli* chromosome.

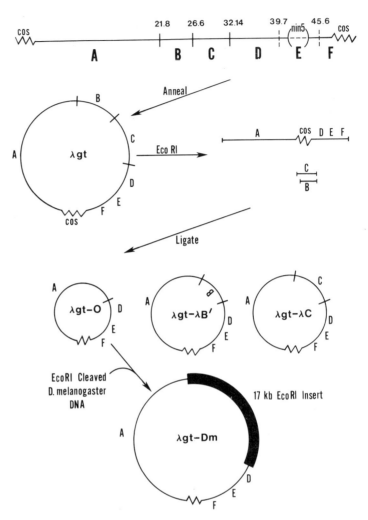

Figure 1.10
Cloning *Drosophila melanogaster* DNA using λgt-0. The construction of λgt-0 from bacteriophage λ is shown. The numbers above the physical map of λ represent the kilobase coordinates. The capital letters below the map represent the 6 *Eco*RI restriction fragments of λ. The broken lines below coordinates 39.7 and 45.6 represent the mutational loss of two *Eco*RI sites. (Thomas et al., Proc. Natl. Acad. Sci. USA 71:4579, 1974.)

arm of λgt. The *nin5* deletion removes 6.1% of the λ genome in a region containing a strong terminator for rightward RNA synthesis. This produces two desirable effects. First, the *nin5* deletion insures the expression of those genes required for cell lysis (S and R); second, the removal of fragments B and C now constitutes a deletion of 29.4% (23.3% + 6.1%) of the λgt genome. As mentioned earlier, λ genomes smaller than 75% cannot be packaged, thus preventing lytic growth and phage production. Therefore only those λgt molecules that incorporate a minimum of 2.2 kb, or a maximum of 16.9 kb, will form plaques after transfection (fig. 1.10).

The generalized transducing (gt) derivatives of λ were just the first in a long line of replacement-type phage cloning vectors. Most of the 100 or so λ vectors currently in use have been modified in a number of ways to make them more efficient and versatile. Some of the more important modifications are listed here.

1. Biological Containment: Due to the potentially hazardous nature of some cloning experiments, there may be a requirement for physical as well as biological containment. Biological containment insures that the propagation of host and vector will be confined to special growth conditions found only in a laboratory setting. By introducing nonsense mutations in the genes required for lytic growth, a number of "safe" λ vectors have been constructed. As shown at the bottom of Fig. 1.9, the λgtWES-λB vector carries amber mutations in the W, E, and S genes, while Charon 4A has amber mutations in gene functions A and B. These mutations prevent phage production in all but a few special strains of *E. coli* K-12 that are capable of suppressing amber mutations.

2. Vector Versatility: As cloning experiments become more sophisticated and demanding, so increases the demand for λ vectors that can accommodate a wider size range of different kinds of restriction fragments. While the original λ vectors were based on the use of *Eco*RI restriction endonuclease, it was inevitable that *Eco*RI sites would fall within, rather than flanking, a gene of interest. Therefore, several different restriction maps of λ were constructed (see Appendix 3) and the appropriate mutations introduced to give λ vectors with broader cloning capabilities. For example, while λgtWES-λB́ has an upper and lower fragment size limit of 15 kb and 1 kb, respectively, Charon 4A has upper and lower limits of 22.2 kb and 8.2 kb, respectively. One of the most versatile of the λ vectors was recently constructed by F. Blattner's group at the University of Wisconsin. As shown in Table 1.1, Charon 30 has a maximum and minimum size limitation of 20.2 kb and 0 kb, respectively, and can be used with eleven different restriction enzymes and restriction enzyme combinations.

Table 1.1
Predicted cloning capacities for Charon 30

ENZYMES		MAXIMUM SIZE	MINIMUM SIZE
*Bam*HI	Alone	19,100	6,100
*Hin*dIII	Alone	11,671	0
*Eco*RI	Alone	17,459	4,459
*Sal*I	Alone	12,175	0
*Xho*I	Alone	11,671	0
*Eco*RI	*Bam*HI	20,243	7,243
*Hin*dIII	*Sal*I	17,048	4,048
*Hin*dIII	*Xho*I	17,308	4,308
*Eco*RI	*Sal*I	18,986	5,986
*Eco*RI	*Xho*I	19,246	6,246
*Sal*I	*Xho*I	12,435	0

The predicted cloning capacities shown in this table are based on a maximum and minimum size for λ of 51 kb and 38 kb, respectively. (Modified from D. L. Rimm et al., Gene 12:301, 1980.)

CHAPTER 1

INTRODUCTION TO RECOMBINANT DNA TECHNOLOGY

3. *In vitro* Packaging: While not a modification of the λ vector itself, *in vitro* packaging of recombinant λ DNA represents an important procedural modification that has greatly enhanced the cloning efficiencies of all λ-based vectors. Due to the large number of recombinant molecules needed to construct a gene bank of a higher eukaryote, plaque-forming efficiencies are required that are orders of magnitude higher than those that can be achieved by transfection. Even under the best conditions, transfection of $CaCl_2$-treated cells of *E. coli* with recombinant molecules will yield only 10^3 to 10^4 plaques per microgram DNA. However, if the products of the ligation reaction are first packaged into phage coats (with tails), plaquing efficiency can be as high as 10^6 plaques per microgram DNA. *In vitro* packaging can be accomplished by adding the ligation products to a cell lysate mixture derived from two genetically distinct λ-lysogenic strains of *E. coli*. Separately, the lysates of each of the lysogenic strains are incapable of packaging λ DNA. However, when mixed together, the lysates can package recombinant and nonrecombinant λ DNA into infectious particles, by *in vitro* genetic complementation. Once the recombinant λ DNA molecules have been packaged, they can then enter the host cell by the natural and more efficient means of absorption and injection.

4. Cosmids: The most novel modification of the λ vector cloning system is the construction of cosmids (*cos* + plasmids). Cosmids combine the best features of the plasmid cloning vector with the maximum cloning capacities of the λ vectors. For example, the cosmid HC79 (6.1 kb) consists of a 1.7 kb DNA fragment containing the *cos* region of λ inserted into a nonessential region of pBR322. Foreign DNA fragments can be cloned into any one of the unique restriction sites on pBR322 and the products of the ligation packaged *in vitro*. Since the λ packaging endonucleases require only a head-full of DNA separated by two *cos* sites, ligation products consisting of long concatamers (from Latin, con = together; catena = chain) of cosmid and stuffer DNA can be packaged. However, as in the case of λ DNA packaging, fragment size requirements must be met. When linked to a 6 kb cosmid, stuffer DNA fragments greater than 31 kb, but less than 45 kb, can be efficiently packaged into infectious phage particles. After infection, however, the host cells are selected for growth on antibiotic-containing medium (e.g., Luria broth containing ampicillin) and the colonies that form are analyzed for the presence of plasmid DNA. As with the λ vector system, cosmids also provide a positive selection for recombinant DNA molecules. Unlike the λ vectors, recombinant cosmids are maintained as plasmids in the host cell instead of in the bacteriophage state and the maximum insert size limitation is 45 kb as opposed to 22 kb for λ.

Most cloning vectors can be categorized according to purpose, host-range, and type of extrachromosomal element involved. The different categories of cloning vectors and some representative examples are summarized in Table 1.2.

Table 1.2
Representative examples of cloning vectors

I. Generalized Cloning Vectors (for the unbiased cloning of genomic and cDNA fragments for genetic and biochemical analysis)
   A. *Escherichia coli* vectors
      1. Plasmids
         a) Insertional inactivation without direct selection for recombinant plasmids
            (1) pACYC184 (Chang and Cohen, J. Bacteriol. 134:1141, 1978)
            (2) pBR329 (Covarrubias and Bolivar, Gene 17:79, 1982)
            (3) pKH47 (Hayashi, Gene 11:109, 1980)
            (4) pUC7, 8, and 9 (Vieira and Messing, Gene 19:259, 1982)
            (5) pBR322 (Bolivar et al., Gene 2:95, 1977)
         b) Insertional inactivation with positive selection for recombinant plasmids
            (1) pHE3 (Hennecke et al., Gene 19:231, 1982)
            (2) pKL1 (Honigman et al., Gene 13:289, 1981)
            (3) pLV59 (O'Connor and Humphreys, Gene 20:219, 1982)
            (4) pNO1523 (Dean, Gene 15:99, 1981)
            (5) pTR262 (Roberts et al., Gene 12:123, 1980)
         c) Replacement vectors with positive selection for recombinant plasmids
            (1) pGJ53 (Hagan and Warren, Gene 19:147, 1982)
      2. Single-stranded bacteriophage vectors
         a) Insertional inactivation without direct selection for recombinant phage
            (1) M13mp7, 8, and 9 (Messing and Vieira, Gene 19:269, 1982)
            (2) fd100 series (Herrmann et al., Mol. Gen. Genet. 177:231, 1980)
      3. Double-stranded bacteriophage vectors
         a) Replacement-type vectors with direct selection for recombinant phage
            (1) λ, Charon 4A (Blattner et al., Science 196:161, 1977)
            (2) λ, Charon 30 (Rimm et al., Gene 12:301, 1980)
            (3) λNM 570-BV1, -BV2 (Klein and Murray, J. Mol. Biol. 133:289, 1979)
         b) Insertional inactivation and replacement vectors with positive selection
            (1) λ, Charon 16A (Blattner et al., Science 196:161, 1977)
      4. Cosmid vectors
         a) pHC79 (Hohn and Collins, Gene 11:291, 1980)
         b) MUA-3 (Meyerowitz et al., Gene 11:271, 1980)
   B. Non-*Escherichia coli* vectors
      1. Plasmids—Broad-host range, gram-negative
         a) pK1230, pK1262 (Bagdasarian et al., Gene 16:237, 1981)
         b) pRK290 (Ditta et al., Proc. Natl. Acad. Sci. USA 77:7347, 1980)
         c) pGV1106, pGV1113 (Leemans et al., Gene 19:361, 1982)
         d) pGSS15 (Hennan et al., Nature 297:80, 1982)
         e) pSa151 (Tait et al., Biotechnology 1:269, 1983)

(*continued*)

CHAPTER 1

INTRODUCTION TO
RECOMBINANT DNA TECHNOLOGY

    2. Plasmids—Gram-positive bacteria
       a) Insertional inactivation without positive selection
         (1) for *B. subtilis*, pHV33 (Michel et al., Gene 12:147, 1980)
         (2) for *B. subtilis* pUB110, pBD9 (Jalanko et al., Gene 14:325, 1981)
         (3) for Streptomyces pJI61 (Thompson et al., Gene 20:51, 1982)
         (4) for Streptococcus pVA838 (Macrina et al., Gene 19:345, 1982)
  C. Eukaryotic cloning vectors
    1. Plasmids
       a) For yeast
         (1) YEp, YRp, and YIp (Botstein et al., Gene 8:17, 1979)
         (2) pRC1, pRC2, and pRC3 (Ferguson et al., Gene 16:191, 1981)
         (3) pYc(CEN3)41 (Clarke and Carbon, Nature 287:504, 1980)
       b) For mammalian cells
         (1) SV40 (Hamer et al., J. Mol. Biol. 112:155, 1977)
         (2) SVGT5 (Mulligan et al., Nature 277:108, 1979)
         (3) pGP (Kushner et al., J. Mol. Appl. Genet. 1:527, 1982)
         (4) pSV-neo (Southern and Berg, J. Mol. Appl. Genet. 1:327, 1982)

II. Specialized Plasmid Cloning Vectors
  A. *E. coli* cloning vectors
    1. Plasmids
       a) Expression vectors
         (1) pKT100, pKT200 (Talmadge and Gilbert, Gene 12:235, 1980)
         (2) pPLa2311 *trp*A1 (Remaut et al., Gene 15:81, 1981)
         (3) pJJS1002 (Sninsky et al., Gene 16:275, 1981)
         (4) pOP203-13, pOP95-15 (Fuller, Gene 19:43, 1982)
         (5) pMOB45 (Bittner and Vapnek, Gene 15:319, 1981)
         (6) pTR865 (Bikel et al., Proc. Natl. Acad. Sci. USA 80:906, 1983)
       b) Promoter cloning vectors
         (1) pGA24, pGA79 (An and Griesen, J. Bacteriol. 140:400, 1979)
         (2) pPV33, pPV501 (West and Rodriguez, Gene 20:291, 1982)
         (3) pEP3012, pEP3015 (Enger-Valk et al., Gene 15:297, 1981)
         (4) pCM1, pCM4, and pCM7 (Close and Rodriguez, Gene 20:305, 1982)
         (5) pKM-1 (deBoer et al., Proc. Natl. Acad. Sci. USA 80:21, 1983)
         (6) pKO-1 (Rosenberg et al., in Promoters: Structure and Function, R. L. Rodriguez and M. J. Chamberlin, eds., Praeger Scientific, New York, 1982)
       c) Terminator cloning vectors
         (1) pEP165 (Enger-Valk et al., Nucl. Acids Res. 9:1973, 1981)
         (2) pKL1, pHA10 (Honigman et al., Gene 13:289, 1981)
         (3) pKO-1 (Rosenberg et al., in Promoters: Structure and Function, R. L. Rodriguez and M. J. Chamberlin, eds., Praeger Scientific, New York, 1982)
         (4) pMC81 (Casadaban and Cohen, J. Mol. Biol., 138:179, 1980)

2. Double-stranded bacteriophage vectors
      a) Promoter-cloning vectors
         (1) Mu-d1 (Casadaban and Cohen, Proc. Natl. Acad. Sci. USA 76:4530, 1979)
         (2) pKO-1 (Rosenberg et al., in Promoters: Structure and Function, R. L. Rodriguez, M. J. Chamberlin, eds., Praeger Scientific, New York, 1982)
      b) Terminator-cloning vector
         (1) pKO-1 (Rosenberg et al., in Promoters: Structure and Function, R. L. Rodriguez and M. J. Chamberlin, eds., Praeger Scientific, New York, 1982)

B. Eukaryotic cloning vectors
   1. Plasmid (and plasmid/viral hybrids)
      a) Expression vectors for yeast
         (1) pFRS series (Hitzeman et al., Nature 293:717, 1981)
      b) Expression vectors for mammalian cells
         (1) pSV529 (Gheyser and Fiers, J. Mol. Appl. Genet. 1:385, 1982)
         (2) Ad-SVR26 (Thummel et al., J. Mol. Appl. Genet. 1:435, 1982)
      c) Promoter/Terminator cloning vectors for mammalian cells
         (1) pSVK (Rosenberg et al., in Promoters: Structure and Function, R. L. Rodriguez and M. J. Chamberlin, eds., Praeger Scientific, New York, 1982)

Cloning Strategies

A number of different strategies can be used to clone a particular DNA fragment or gene. Choosing the right strategy is an important decision and should take into account factors such as (1) the nature of the target DNA, (2) the availability of screening methods, and (3) the selection schemes. Generally speaking, cloning strategies can be divided into two basic categories—those that involve the cloning of specific fragments of DNA and those which involve the construction of "gene banks" (or libraries).

In some instances, particularly with the larger genomes of the higher eukaryotes, it is necessary to clone the maximum number of different restriction fragments (gene bank) to obtain the gene of interest at a reasonable frequency. For example, the human β-hemoglobin gene is approximately 1.5 kb, while the human haploid genome is about 3×10^6 kb. How many different restriction fragments, 1.5 kb in length, have to be cloned to ensure (with 99% certainty) that the β-globin gene will be represented at least once among the recombinant transformants? The answer can be obtained using the following mathematical expression developed by Clarke and Carbon:

$N = \ln(1 - P)/\ln(1 - f)$;

where

   N = number of recombinant transformants
   P = the probability that a given sequence will be represented among the recombinants
   f = the average restriction fragment size as a fraction of the total genome size.

For this example, the answer (N) can be calculated as follows:

$$N = \frac{\ln(1 - 0.99)}{\ln[1 - (1.5 \times 10^0/3 \times 10^6)]} = 9.2 \times 10^6$$

Using a vector without a means for selecting only recombinant transformants (positive selection) would make isolating and characterizing 9 million clones extremely tedious and time consuming. Furthermore, when the efficiencies of ligation and transformation are taken into account, the likelihood of obtaining this many recombinants in one experiment is very low. However, if a λ vector such as Charon 4A is used to clone human DNA fragments with an average size of 15 kb, the number of recombinants needed for the gene bank can be reduced to 9×10^5 clones. If the cosmid pHC79 is used to clone fragments with an average size of 40 kb, the number of recombinants is further reduced to about 3.5×10^5. Using the appropriate probe (see Chapter 8) and the high-density plaque and colony hydridization procedures currently available, $3-9 \times 10^5$ plaques or colonies can be screened for the presence of the β-globin gene more quickly and conveniently than if a plasmid vector were used. As a general rule, the λ vectors and cosmids are used to clone large DNA fragments and for the construction of gene banks, while the plasmid vectors are used to clone smaller DNAs (<15 kb) for which a means of detection is readily available.

The nature of the DNA to be cloned is another important factor to consider when choosing a cloning strategy. For example, a gene such as human β-globin can be cloned in two forms: the genomic copy and the cDNA copy. A cDNA copy made from mature (processed) β-globin mRNA by using reverse transcriptase, will differ from the genomic copy. The cDNA copy will lack the regulatory signals located at the 5'-nontranscribed-region of the gene, as well as the introns. Introns (untranslated intervening sequences) are characteristic of many eukaryotic genes and are removed from the mRNA before translation begins. As a consequence, a cDNA copy of a gene will be small (<2–3 kb) relative to the maximum cloning capacities for λ and the cosmids. Therefore, plasmid vectors such as pBR322 are commonly used to clone cDNA fragments. Since introns have yet to be found in prokaryotes, the distinction drawn here between genomic and cDNA copies of the same gene has little bearing on the cloning of prokaryotic DNA.

In the lexicon of recombinant DNA terminology, the distinction made between the terms *selection* and *screen* is an important one. Selection involves the use of a genetic or biochemical "trick" to eliminate (i.e., inhibit the growth of) all cells except those harboring the desired recombinant plasmid. A screen, on the other hand, only provides a means for distinguishing the desired clone from the surrounding recombinant and nonrecombinant transformants.

A selection scheme may be based on some property of the plasmid vector or it may involve some special feature of the cloned fragment itself. The best examples of vectors that allow positive selection for recombinant molecules are the replacement-type λ vectors mentioned earlier. An example of a plasmid vector that provides a positive selection scheme is pTR262 (Table 1.2). This plasmid derivative of pBR322 contains the λ repressor gene and the rightward operator and promoter ($O_R P_R$) inserted next to the structural portion of the tetracycline resistance gene. This arrangement puts the expression of the tetracycline resistance gene under the control of transcription initiating from P_R. Transcription from P_R is itself controlled by the λ repressor protein acting on O_R to block expression of tetracycline resistance. Since the cI gene contains a unique *Hind*III site, this gene can be insertionally inactivated by cloning DNA fragments in the *Hind*III site. In the absence of a functional repressor, the tetracycline resistance gene is expressed. Therefore, only bacterial cells transformed with *Hind*III-generated recombinant molecules will grow on media containing tetracycline.

If the target DNA fragment contains a gene that complements a host mutation or confers some new phenotype to the host cell, positive selection of recombinant transformants can be achieved without the use of specialized vectors. Using the plasmid ColE1, Carbon's group was able to construct a gene bank of *E. coli* DNA and select for a number of *E. coli* genes by the ability of the cloned fragments to complement different host mutations. Similar results have been obtained using yeast DNA, although the number of yeast genes capable of complementing *E. coli* defects is relatively small.

An example of a selection scheme based on the ability of a cloned fragment to confer a new phenotype to the host is provided by the work of Chang et al. In these experiments, a mouse gene bank was constructed from cDNA made from total mouse mRNA. On the basis of the resistance of mouse dihydrofolate reductase to the antibiotic trimethoprim, the mouse dihydrofolate reductase gene was isolated by growing the transformants on medium containing trimethoprim at a level that inhibits the analogous enzyme in *E. coli*. The DNA fragment containing the mouse dihydrofolate reductase gene conferred a new phenotype (trimethoprim resistance) to the *E. coli* host.

Finally, a selection scheme can be achieved based on the modification of vector or target DNA prior to ligation. These modifications are designed to force the ligation of vector to target DNA in the reaction mixture, thus preventing reclosure of the vector on itself. As will be discussed in detail in Chapter 6, one way to optimize the formation of recombinant molecules employs the enzyme alkaline phosphatase to remove the phosphate groups from cohesive ends generated by restriction endonuclease. Since these phosphate groups are essential for ligation (see Chapter 6), the treated vector cannot recircularize. However, if untreated target DNA is added to the reaction mixture, partial ligation of the two DNAs can occur. Once in the cell, the two remaining gaps will be sealed during the first round of plasmid DNA replication.

Two screening methods have already been presented in the discussion of plasmid and M13 cloning vectors. Insertional inactivation of a plasmid-encoded gene changes the genotype of the plasmid so that recombinant versus nonrecombinant transformants can be distinguished on the basis of altered phenotype. As shown in Fig. 1.3, the insertion of a 1,600 bp DNA fragment into the *Bam*HI site of pBR322 destroys the function of the tetracycline resistance gene. Tetracycline-sensitive transformants can be differentiated from tetracycline-resistant cells by deter-

CHAPTER 1

INTRODUCTION TO RECOMBINANT DNA TECHNOLOGY

CHAPTER 1

INTRODUCTION TO RECOMBINANT DNA TECHNOLOGY

mining their ability to grow on tetracycline-containing media (see Chapter 8). With the M13 and pUC vectors, this transfer of cells to a special medium is unnecessary since recombinant transformants can be visualized directly on the plates as blue plaques (or colonies).

A number of screening methods are available to identify specific genes or DNA fragments. For example, the alkaline phosphatase gene (*phoA*) was cloned using a *phoA*⁻ strain of *E. coli* and the indicator dye XP (5-bromo-4-chloro-3-indolylphosphate-p-toluidine). When transformants were grown on media containing XP, those that possessed alkaline phosphatase activity appeared as blue colonies. As in the case of Xgal, cleavage of XP by alkaline phosphatase produces a chromophore that turns the cells blue.

Other specific screening methods rely on base complementary or antigen/antibody interaction. When RNA (cDNA) or antibody specific for a certain gene or gene product is labeled with radioisotopes, it can be used as probes to detect the desired clone among the total population of transformed cells (see Chapter 8). Under the appropriate conditions, cDNA and RNA probes can seek out and hybridize to complementary sequences on the cloned fragment. The resulting homoduplex can then be visualized by autoradiography.

If the screening method is based on the expression of antigen or cross-reacting material in the transformant, special consideration must be given to the site of insertion of the target DNA in the plasmid vector. In most cases, cDNA copies of eukaryotic genes are inserted downstream of plasmid promoters, in the path of transcription. If the fragment is inserted within a plasmid-encoded gene, in the correct orientation and reading frame, the cell will synthesize a fusion polypeptide that will react with antibodies made against the gene product.

One of the best examples of a cloning experiment that employs many of the techniques just described is the cloning of the rat preproinsulin gene in *E. coli* (fig. 1.11). Preproinsulin mRNA was isolated from a rat insulinoma and used to synthesize 0.25 microgram (µg) of cDNA. Using the enzyme terminal transferase (see Appendix 4), approximately 15 deoxycytidine triphosphate (dCTP) residues were added to the ends of the cDNA fragments. In a separate reaction, dGTP residues were added to the ends of *Pst*I cleaved pBR322 DNA. Due to the extensive base homology between the C-tailed ends of the cDNA and the G-tailed ends of the vector, when added together, they annealed to form stable recombinant molecules. Upon transformation of *E. coli*, 2,355 tetracycline-resistant transformants were obtained. Prior to the C-tailing reaction, a portion of the cDNA was removed and labeled with ^{32}P to serve as a hybridization probe. Using the colony hybridization technique (Chapter 8) and other analytical procedures, 48 clones were obtained that contained the rat insulin sequence. One clone harbored a plasmid, pI47, that contained a *Pst*I fragment complementary to the cDNA sequence. When anti-insulin antibody was used to assay this clone for the presence of material that was reactive with anti-insulin antibodies, a positive response was obtained. Furthermore, just as β-lactamase, the product of the ampicillin resistance gene, is secreted into the periplasmic space (the re-

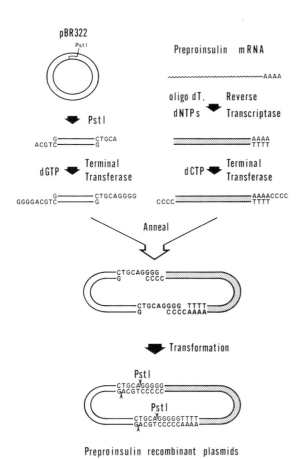

Figure 1.11
Cloning of the rat preproinsulin gene in *E. coli*. (Villa-Komaroff et al., Proc. Natl. Acad. Sci. USA 75:3727, 1978.)

gion between the cell wall and membrane), so was the cross-reacting material. When the structure of the protein and the nucleotide sequence of the cloned insert were determined for pI47, the following observations were made:

1. The G-C tailing reaction regenerated *Pst*I sites at the ends of the cDNA insert.
2. The cDNA insert in pI47 contained the coding sequence for preproinsulin and the orientation of the insert was the same as β-lactamase (i.e., β-lactamase → preproinsulin →).
3. The G-C tailing reaction had added the correct number of bases in pI47 to put the insert in the same reading frame as β-lactamase.
4. A fusion polypeptide consisting of the first 184 amino acids of β-lactamase and the entire 107 amino acids of preproinsulin was synthesized in *E. coli*.
5. The 23 amino acid signal peptides (required for secretion) for both β-lactamase and preproinsulin were removed upon entry of the two proteins into the periplasmic space. Therefore, a bacterial strain that secreted rat proinsulin had been constructed.

CHAPTER 1

INTRODUCTION TO RECOMBINANT DNA TECHNOLOGY

The experiment just described is clearly beyond the scope and intent of this lab manual; however, it does serve to illustrate how even the most advanced cloning experiment must rely on the basic components of recombinant DNA technology, and may involve both simple and sophisticated cloning strategies. Furthermore, this experiment demonstrates how success in cloning requires sound experimental design, an understanding of the biology of the system, and technical expertise. While the experiments set forth in this manual may seem modest in comparison to the one just described, they are designed to give each student the background needed to proceed further.

Preface to the Experiments

The fifteen exercises contained in Chapters 2 through 9 are designed to expose the student to the basic concepts and techniques of recombinant DNA technology and its application to the genetic analysis of bacterial genes. The specific object of these exercises is to clone genes from this histidine and arabinose operons of *E. coli*. The general purpose and content of each chapter is outlined below.

Chapter 2: The principles of bacterial growth and expression of antibiotic resistance are presented. Students are provided with the bacterial strains that will be used in all subsequent experiments. The exercises presented in this chapter instruct the student in the storage of bacterial strains and methods for determining growth rate. The phenotypes of both auxotrophic and plasmid-containing strains will be determined.

Chapter 3: Principles and techniques for the isolation of DNA are presented in this chapter. Using the bacterial strains provided in Chapter 2, *E. coli* and plasmid DNAs will be purified and used for cloning in subsequent experiments. A method for the rapid isolation of plasmid DNA is also presented.

Chapter 4: *E. coli* and plasmid DNA prepared in Chapter 3 will be digested with different restriction endonucleases. Samples of chromosomal and plasmid DNA will be digested and prepared for ligation in Chapter 6. The introduction to the exercises in this chapter discusses host restriction modification systems and the type-II restriction endonucleases.

Chapter 5: This chapter discusses the principles and practical aspects of DNA gel electrophoresis. DNAs digested with restriction enzymes in Chapter 4 will be analyzed by agarose gel electrophoresis. Methods for estimating the size of restriction fragments are discussed.

Chapter 6: The theoretical and practical aspects of ligation are discussed in the introduction to this chapter. A method for predicting the products of the ligation reaction is given. DNAs prepared and analyzed in previous exercises are ligated and prepared for transformation.

Chapter 7: A detailed discussion of natural and artificial transformation systems is presented. Competent cells of the auxotrophic recipient strains are prepared for transformation. The recombinant DNA molecules prepared by *in vitro* ligation in Chapter 6 will be used to transform the appropriate competent strains of *E. coli*.

Chapter 8: Several methods for detecting and characterizing recombinant transformants are discussed. Strategies for mapping genes on cloned DNA fragments are also presented. Histidine and arabinose positive (His^+ and Ara^+) transformants obtained from Chapter 7 will be characterized with respect to phenotype. Restriction enzyme cleavage maps of His^+ and Ara^+ plasmids will be constructed.

Chapter 9: The final chapter of this manual is devoted to a discussion of regulatory DNA sequences (e.g., promoters and terminators) and the use of specialized plasmid vectors for their cloning. Promoter-probe plasmid vectors prepared in earlier experiments will be used to clone the promoter regions of the His^+ and Ara^+ plasmids isolated and characterized in Chapter 8.

Finally, additional information on genetic, physiological, and molecular aspects of the histidine and arabinose operons can be obtained from the following selected readings.

Histidine Operon

Artz, S. W., and Broach, J. R. 1975. Histidine regulation in *Salmonella typhimurium:* An activator-attenuator model of gene regulation. Proc. Natl. Acad. Sci. USA 75:3453–3457.

Barnes, W. M. 1978. DNA sequence from the histidine operon control region: Seven histidine codons in a row. Proc. Natl. Acad. Sci. USA 75:4281–4285.

Bruni, C. B., Musti, A. M., Frunzio, R., and Blasi, F. 1980. Structural and physiological studies of the *Escherichia coli* histidine operon inserted into plasmid vectors. J. Bacteriol. 142:32–42.

DiNocera, P. P., Blasi, F., DiLauro, R., Frunzio, R., and Bruni, C. B. 1978. Nucleotide sequence of the attenuator region of the histidine operon of *Escherichia coli* K-12. Proc. Natl. Acad. Sci. USA 75:4276–4280.

Frunzio, R., Bruni, C. B., and Blasi, F. 1981. *In vivo* and *in vitro* detection of the leader RNA of the histidine operon of *Escherichia coli* K-12. Proc. Natl. Acad. Sci. USA 78:2767–2771.

Johnston, H. M., Barnes, W. M., Chumley, F. G., Bossi, L., and Roth, J. R. 1980. Model for regulation of the histidine operon of *Salmonella*. Proc. Natl. Acad. Sci. USA 77:508–512.

Rodriguez, R. L., West, R. W., Tait, R. C., Jaynes, J. M., and Shanmugan, K. T. 1981. Isolation and characterization of the *his*G and *his*D genes of *Klebsiella pneumoniae*. Gene 16:317–320.

Arabinose Operon

Greenfield, L., Boone, T., and Wilcox, G. 1978. DNA sequence of the *ara* BAD promoter in *Escherichia coli* B/r. Proc. Natl. Acad. Sci. USA 75:4724–4728.

Kaplan, D. A., Greenfield, L., Boone, T., and Wilcox, G. 1978. Hybrid plasmids containing the *ara* BAD genes of *Escherichia coli* B/r. Gene 3:177–187.

CHAPTER 1

INTRODUCTION TO RECOMBINANT DNA TECHNOLOGY

Lee, N. L., Gielow, W. O., and Wallace, R. G. 1981. Mechanism of *ara* C autoregulation and the domains of two overlapping promoters P_C and P_{BAD}, in the L-arabinose regulatory region of *Escherichia coli*. Proc. Natl. Acad. Sci. USA 78:752–756.

Miyada, C. G., Soberón, X., Itakura, K., and Wilcox, G. 1982. The use of synthetic oligodeoxyribonucleotides to produce specific deletions in the *ara* BAD promoter of *Escherichia coli* B/r. Gene 17:167–177.

Ogden, S., Haggerty, D., Stoner, C. M., Kolodrubetz, D., and Scheif, R. 1980. The *Escherichia coli* L-arabinose operon: Binding sites of the regulatory proteins and a mechanism of positive and negative regulation. 1980. Proc. Natl. Acad. Sci. USA 77:3346–3350.

Wallace, L. J., and Wilcox, G. 1979. Regulation of the L-arabinose operon in strains of *Escherichia coli* containing ColE1-*ara* hybrid plasmids. Molec. Gen. Genet. 173:323–331.

Wilcox, G., Al-Zarban, S., Cass, L. G., Clarke, P., Heffernan, L., Horwitz, A. H., and Miyada, C. G. DNA sequence analysis of mutants in the *ara* BAD and *ara* C promoters. In Promoters: Structure and Function, R. L. Rodriguez and M. J. Chamberlin, eds., Praeger Scientific, New York, 1982.

West, R. W., Kline, E. L., Horwitz, A. H., Wilcox, G., and Rodriguez, R. L. Cloning and analysis of inducible *Escherichia coli* promoters using promoter-probe plasmid vectors. In Promoters: Structure and Function, R. L. Rodriguez and M. J. Chamberlin, eds., Praeger Scientific, New York, 1982.

References

References to most of the concepts, techniques, and experiments discussed in Chapter 1 can be found in the following sources.

Grossman, L., and Moldave, K. (eds.). 1980. Methods in Enzymology. Volume 65, Nucleic Acids, Part I. Academic Press, New York.

Maniatis, T., Fritsch, E. F., and Sambrook, J. 1982. Molecular Cloning: A Laboratory Manual. Cold Spring Harbor Laboratory, New York.

Old, R. W., and Primrose, S. B. 1981. Principles of Gene Manipulation: An Introduction to Genetic Engineering. University of California Press, Berkeley.

Wu, R. (ed.). 1979. Methods in Enzymology. Volume 68, Recombinant DNA. Academic Press, New York.

CHAPTER 2

Microbiological Techniques

The bacterium *Escherichia coli* can be easily grown and maintained in liquid medium or on medium that has been solidified by the addition of agar. The particular nutritional requirements and the antibiotic resistance pattern of a specific strain serve as a "fingerprint" that can help to identify the strain. For example, indicator dyes such as tetrazolium or X-gal and minimal agar media can be used to verify genetic markers involving defects in metabolic functions.

During the growth of a bacterial culture, two problems may arise that affect the purity of the culture: the contamination of the culture with other bacteria, and genetic heterogeneity. In order to minimize contamination, all materials must be sterilized prior to contact with the culture. For media and dilution buffers, this can be achieved by either autoclaving or passing the solution through a filter designed to remove bacteria. Glassware may be sterilized in an autoclave or by exposure to dry heat. Plastic materials such as disposable petri dishes can be treated with gaseous sterilizing agents. In order to prevent airborne contamination of a culture by bacterial and fungal spores, culture flasks and tubes are sealed with sterile cotton plugs or caps. As a final safeguard, aseptic techniques are always used during the manipulation of bacterial cultures.

The second major source of heterogeneity in a bacterial culture is the appearance of altered derivatives of the original bacterial strain. These derivatives may be the result of spontaneous mutations that change the original phenotypic markers. In the case of plasmid-borne markers, heterogeneity may be the result of the loss or segregation of the plasmid from the dividing bacteria. Segregation is an important aspect when dealing with plasmid-bearing strains, since many plasmids are known to segregate from the host cell at fairly high frequencies.

Antibiotics inhibit bacterial growth in either a bacteriostatic or a bacteriocidal manner. Drugs that result in the inhibition of cell growth without directly causing cell death are bacteriostatic, while those that cause cell

Antibiotic Resistance

CHAPTER 2
MICROBIOLOGICAL TECHNIQUES

lysis or death are bacteriocidal in nature. The bacteriostatic or bacteriocidal effect of a particular antibiotic depends on the conditions under which bacteria are exposed to the drug. Exposure of E. coli to low levels of tetracycline (Tc) will inhibit cell growth by binding to the ribosome and interfering with protein synthesis. However, since the binding of Tc to the ribosome is reversible, the inhibited cells can be diluted into drug-free medium and normal cell growth will resume. The reversible nature of this inhibition indicates that the Tc is functioning in a bacteriostatic manner. At higher concentrations of Tc, other inhibitory effects of the drug become apparent and the inhibition becomes irreversible. Thus, at high concentrations, Tc may function in a bacteriocidal manner.

Bacterial resistance to antibiotics can be generated by one of three mechanisms: by alteration of the target site with which the antibiotic interacts, by preventing entry of the drug into the cell, or by enzymes that modify or inactivate the drug. For example, resistance to the protein synthesis inhibitor streptomycin (Sm) is often caused by the synthesis of an altered ribosomal protein that no longer binds Sm, thereby preventing the inhibitory action of Sm. In the case of resistance to Tc, membrane permeability to Tc is decreased by the synthesis of membrane proteins that block the entry of Tc into the cell. Resistance to penicillin is often caused by the secretion of the enzyme β-lactamase, which is able to cleave and inactivate the antibiotic.

The lowest antibiotic concentration that inhibits cell growth is often referred to as the minimum inhibitory concentration (MIC). Cells resistant to one concentration of antibiotic may not grow if the antibiotic concentration is doubled. With many bacteriostatic antibiotics, such as Tc, the bacterial colony size will become smaller as the concentration of antibiotic in the medium approaches the MIC. When cells resistant to a Tc concentration of 80 μg/ml are mixed with cells resistant to a Tc concentration of 30 μg/ml and the mixture is plated on an agar plate containing 30 μg/ml Tc, two colony types will be observed, as illustrated in Fig. 2.1a. The cells resistant to the higher Tc concentration will appear as large colonies, while the less resistant cells form smaller colonies, sometimes referred to as microcolonies. If the two colony types are plated on medium containing 15 μg/ml Tc, their respective colony morphologies will be indistinguishable.

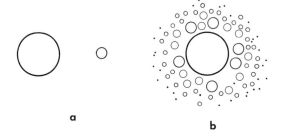

Figure 2.1
Unusual colony morphology commonly observed on agar plates containing antibiotics. (a) A comparison of normal colony size relative to a microcolony. (b) An example of a bacterial colony surrounded by "feeder" colonies.

Ampicillin, a derivative of penicillin, is a broad spectrum bacteriocidal antibiotic that inhibits the synthesis of the bacterial cell wall and causes cell lysis. Metabolically inactive cells or those in the stationary phase of cell growth are unaffected by the drug. When a mixture of ampicillin-resistant and ampicillin-sensitive cells are plated on agar plates containing ampicillin, the resistant cells appear as large colonies surrounded by a zone of microcolonies of varying sizes (fig. 2.1b). The β-lactamase secreted by the resistant colonies inactivates the ampicillin in a zone around the resistant colony, allowing the slow growth of ampicillin-sensitive cells in the zone of reduced ampicillin concentration. Further purification of these microcolonies reveals that they are completely sensitive to ampicillin. These microcolonies are often called "feeder" colonies since they grow only in the zone of reduced ampicillin concentration surrounding a colony secreting β-lactamase.

Bacterial Cell Growth

Under conditions of optimal growth, bacteria may reproduce by binary fission at an average rate of once every twenty minutes. However, this rate can be maintained only for a short period. When bacteria are inoculated into a culture medium, they pass through the following periods of growth:

1. An initial lag phase, during which there is little or no multiplication of the bacteria.
2. The logarithmic period, or period when the organisms grow at a maximum rate, increasing by geometric progression as a result of reproduction by fission. At this time, if the logarithms of the number of bacteria present are plotted against time, a straight line is obtained.
3. The stationary period, or period of the plateau, which may be plotted on a growth curve. During this time the bacteria are dying as fast as they are dividing.
4. The period of decline and death, when the growth rate declines steadily and cell death exceeds the rate of cell division.

Most recombinant DNA experiments are performed with bacterial cultures in the log phase of cell growth. Therefore, the investigator must know the concentration of bacterial cells per ml of culture and the approximate rate of growth. There are two basic methods of determining the concentration of bacterial cells in a culture: by total count and by viable count. The total count of bacteria is determined by counting the number of bacteria using a Petroff-Hauser chamber with a phase contrast microscope. The total count includes both viable and nonviable bacteria. The viable count, on the other hand, is determined by spreading appropriate dilutions of a culture on solid medium and allowing bacterial colonies to form. Since each viable cell will produce one colony, the concentration of bacteria in the original culture can be readily calculated. However, the time required for a single cell to produce a visible colony (8 to 16 hours) renders this method impractical for routine determination of cell concentration. Therefore, it is commonplace to relate viable counts to turbidity, a more convenient method for measuring cell density.

CHAPTER 2
MICROBIOLOGICAL TECHNIQUES

The turbidity of a culture is a function of the cell mass/unit volume of culture. The turbidity is due to light scattering, and is best measured at wavelengths where the ratio of absorbance to light scattering is low. A convenient wavelength range for this purpose is between 450 and 570 mμ, and changes in cell mass can be followed by measuring the turbidity of the culture with a spectrophotometer or colorimeter. Since different strains have different cell sizes, and cell size varies with physiological conditions, the turbidity of a culture can only be used as an estimate of cell concentration after turbidity has been calibrated against a direct count of viable cells. Such a calibration is reliable only for the strain on which it was first estimated, and when the culture is in exponential growth phase under specified culture conditions. By combining a plot of turbidity vs. time with a plot of viable counts vs. time, a standard curve can be constructed (fig. 2.2) relating the two parameters, thus giving the investigator a rapid means for measuring cell concentration.

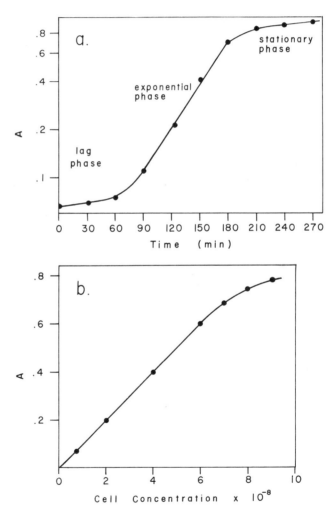

Figure 2.2
(a) Bacterial cell growth plotted as a function of the log of absorbance (A) versus time of incubation and (b) a standard curve relating absorbance (A) to cell concentration.

References

Miller, J. H. Experiments in Molecular Genetics. Cold Spring Harbor Laboratory, New York, 1976.

Stanier, R. Y., Adelberg, E. A., and Ingraham, J. L. The Microbial World. Prentice-Hall, New Jersey, 1976.

EXERCISE 1

Aseptic Manipulation and Storage of Bacterial Cultures

Materials

Bacterial Strains:

Strain #	Designation	Genotype
1	AB257	E. coli K-12 − F⁻, metB, rel-1, PO2B.
2	LA6	E. coli K-12 − F⁺ araΔ766, lac, gal, pro, thi, hsdR, hsdM.
3	BE42	E. coli K-12 − F⁻, hisD3153, lac proΔXIII, thi, gnd.
4	RR1	E. coli K-12 − F⁻, ara (leaky) lac gal, leu, pro, thi, hsdR, hsdM.
5	RR1(pBR329)	Same as RR1 plus, Ap^r, Tc^r, Cm^r.
6	RR1(pPV33)	Same as RR1 plus, Ap^r.
7	RR1(pPV501)	Same as RR1 plus, Ap^r.

Luria Broth (LB):

50 ml in a 250 ml screw-cap Erlenmeyer flask.

7 agar (0.7%) stabs and 7 agar (1.5%) slants.

30 agar plates.

3 agar plates each containing: 20 µg/ml of either ampicillin (Ap), tetracycline (Tc) or chloramphenicol (Cm).

1 agar plate containing 20 µg/ml of Ap, Tc, and Cm.

M9 Minimal Medium (M9 min):

3 agar plates supplemented as follows:

Plate I (M9 min. plus 0.2% glucose, 42 µg/ml histidine, 25 µg/ml methionine, 166 µg/ml proline, 41 µg/ml leucine, and 0.2 µg/ml thiamine hydrochloride (B_1).

Plate II (M9 min. plus 0.2% glucose, 25 µg/ml methionine, 166 µg/ml proline, 41 µg/ml leucine, and 0.2 µg/ml thiamine hydrochloride (B_1).

Plate III (M9 min. plus 0.2% arabinose, 42 µg/ml histidine, 25 µg/ml methionine, 166 µg/ml proline, 41 µg/ml leucine, and 0.2 µg/ml thiamine hydrochloride (B_1).

λ Diluent (λ dil): Sterile, 10 mM Tric-HCl pH 7.6, 50 mM NaCl and 0.01% gelatin.

Protocol

EXERCISE 1

ASEPTIC MANIPULATION AND
STORAGE OF BACTERIAL CULTURES

DAY 1

Each group will receive seven bacterial strains in Luria Broth culture at a cell density of about 10^8–10^9 cells/ml. It will be your responsibility to maintain the purity and viability of each strain. This can be accomplished using the following methods:

Preparation of Stabs

Airtight stabs provide a convenient method of maintaining stocks of *Escherichia coli* at room temperature for about a year (although this varies from strain to strain). For particularly unstable strains, such as Hfr strains, which tend to revert to F^+, storing at $-20°C$ in 20% glycerol is recommended. If the equipment is available, freeze drying is the method of choice, as freeze-dried cultures retain their viability indefinitely.

1. Label each of the 7 stabs with the appropriate strain designation, date, and your initials. Flame the inoculation needle, dip it into the broth culture, then stab it two or three times deeply into the agar. Screw cap tightly and seal with a strip of parafilm. Remember, when using a loop or needle, the wire should be held in the flame for several seconds until the wire glows, incinerating any contaminating organisms. The sterile loop is then dipped into a bacterial culture and used to inoculate the sterile medium. The loop may also be touched to a colony growing on an agar plate to obtain an inoculum; however, if it is not cooled prior to touching the colony, the hot wire will melt the agar medium and kill many of the bacteria. A hot loop may be cooled by touching it to a sterile surface, such as a region of the agar plate that contains no bacterial colonies.

Preparation of Slants

Slant cultures are also a convenient source of inoculum for strains frequently used in the laboratory. After preparation, they can be left at room temperature for about a month. For longer periods the slants can be sealed with parafilm and stored in the refrigerator. Do not rely on this method to store strains for periods longer than 2 to 3 months. During cold storage, some strains lose their viability and bacterial plasmids may be lost through segregation.

1. Flame your inoculation loop and dip it into the liquid culture. Transfer a droplet of the culture onto the surface of a LB agar slant and spread it from the bottom to the top. Flame the tube opening and label with the strain designation, date, and your initials.

Isolation of Single Colonies

Streaking out a bacterial strain on antibiotic or minimal media provides a rapid method for obtaining isolated colonies for subsequent phenotypic characterization. Streak out the strain RR1 (pBR329) to verify the drug resistance phenotype.

EXERCISE 1

ASEPTIC MANIPULATION AND
STORAGE OF BACTERIAL CULTURES

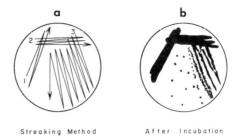

Streaking Method   After Incubation

Figure 2.3
Method for isolating single colonies on agar plates.

1. Dip a sterile inoculation loop into the liquid culture of RR1(pBR329) provided.

2. Touch the droplet to the surface of a LB agar plate containing 20 µg/ml Ap, Tc, and Cm, at position 1 shown in Fig. 2.3a. Draw the droplet across the surface of the plate as indicated.

3. Flame the loop again and place it at position 2 and streak it across the plate as indicated. Repeat the streaking at position 3. As the loop is drawn across the surface of the plate, the cells are spread out and after incubation at 37°C for 12 to 16 hours, colonies will form on the plate in a pattern similar to the one shown in Fig. 2.3b.

4. After colonies have formed, seal the plate with tape and store in the refrigerator for later use in Chapter 3, Exercise 6.

Note: Agar plates should always be incubated face down to prevent: (a) the build-up of excess moisture which will cause the bacterial colonies to diffuse over the surface of the plate to form confluent areas of cell growth, and (b) dessication after prolonged periods of incubation (3 days and longer at 37°C).

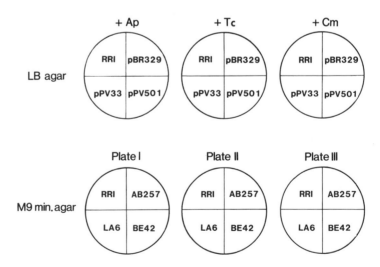

Figure 2.4
Sectored agar plates for verifying phenotypic markers.

Verifying Phenotypic Markers
1. With a felt pen or china marker, sector the antibiotic-containing LB agar and M9 min. agar plates into 4 equal quadrants.
2. Using a sterile inoculation needle, streak the seven strains as shown in Fig. 2.4.
3. Incubate the streaked plates at 37°C for 12 to 16 hours. The M9 min. agar plates will need to be incubated for 36 to 48 hours. Record the results of this experiment as (+) or (−) growth for each quadrant.

DAY 2
1. Record the results of all bacterial strain streakings and phenotypic characterizations.

Preparing a Standard Curve for Bacterial Cell Growth
The number of viable cells/ml in a given culture can be estimated by diluting the culture appropriately and plating out an aliquot on solid medium. Generally dilution is achieved in 100 fold, 10 fold, and 5 fold steps, and 0.1 ml aliquots are spread onto surface of the plates with a bent glass rod. A dilution of the original culture should be selected that will give from 50 to 300 colonies per plate. Low counts are statistically unreliable and plates with too many colonies are difficult to count.

1. Prepare an inoculum of either LA6, BE42, or RR1, 2 to 3 hours before the monitoring of cell growth begins.
2. Using a Bunsen burner, pass a 0.2 ml pipette through the flame to insure that the outer surface of the pipette is sterile. Remove the cap from the tube containing the LB culture of the bacterial strain. Pass the mouth of the tube through the flame, insert the flamed pipette into the culture, and remove the desired 0.1 ml. Flame the mouth of the culture tube again and replace the cap. Remember that while the culture is exposed directly to the air, it is also exposed to an excellent source of contamination. Although it may sound as if three or more hands are required to perform this operation, the small finger of the hand holding the pipette can be used to cap and uncap the tube.
3. Remove the cap or cotton plug from the sterile culture flask containing the 50 ml of LB. Flame the mouth of the flask, then add the 0.1 ml of inoculum from the pipette. Again flame the mouth of the flask, and replace the cap. Incubate the culture at 37°C with shaking for 2 to 3 hours or until the culture reaches an approximate cell density of $1-2 \times 10^7$ cells/ml.
4. After this period of incubation, assume that the 50 ml bacterial culture is approximately 2×10^7 cells/ml. At 30-minute intervals take 0.05 ml of culture and make the following dilutions in λ dil (mix thoroughly before making the next dilution):

T = 0 min	30 min	60 min	90 min	120 min
df = 10^4	2×10^4	2×10^5	5×10^5	10^6

EXERCISE 1

ASEPTIC MANIPULATION AND STORAGE OF BACTERIAL CULTURES

EXERCISE 1

ASEPTIC MANIPULATION AND
STORAGE OF BACTERIAL CULTURES

Use the following information to construct the standard growth curve:

cells/ml = (number of colonies/plate) × (dilution factor) × (volume correction)

where

$$\text{volume correction} = \frac{1}{\text{volume plated (ml)}}$$ and

$$\text{dilution factor (df)} = \frac{\text{original conc.}}{\text{desired conc.}}$$

For example,

$$\frac{2 \times 10^7 \text{ cells/ml}}{2 \times 10^3 \text{ cells/ml}} = \text{df of } 10^4.$$

A dilution factor of 10^4 can be obtained with the following series of dilution: 0.05 ml of culture in 4.95 of λ dil (10^2 df); 0.05 ml of the 10^2 df in 4.95 ml of λ dil (10^4 df). From your final (10^4) dilution, spread a plate with 0.1 ml. Consequently, your volume correction (1/0.1 ml) is 10. This volume will be added directly to the surface of 2 LB agar plates and spread evenly with a glass spreader. About 100 colonies should develop on this plate. You should also plate 10 fold less and 10 fold more than the "ideal" dilution for a total of 6 plates/time interval. Place culture back in shaking incubator quickly. Incubate the plates at 37°C for 18 to 24 hours to allow formation of colonies.

Precautions

1. Label all dilution tubes with the dilution factor.

2. Label the bottoms of all plates with df, volume, correction, date, and initials.

3. Use sterile pipet for each dilution.

4. Make sure the surface of each plate is dry before and after spreading.

5. The glass spreader should be dipped in ethanol, then flamed prior to each use to prevent contamination of plates.

At the same 30-minute intervals used for the viable count determination, take 5 ml of culture and determine the turbity (#54 Green filter, 500–570 nm) using the Klett-Summerson colorimeter. Perform this maneuver quickly since cell growth rate slows when the culture is out of the incubator. Discard the 5 ml sample and continue the incubation.

After counting the colonies and determining the number of cells/ml at each time point, construct curves of cells/ml vs. time, Klett units vs. time, and Klett units vs. cells/ml.

CHAPTER 3

DNA Isolation

One of the first steps in the *in vitro* manipulation of DNA involves the isolation of both the vector DNA and the DNA to be cloned (fig. 3.1). DNA does not exist as a free molecule in the cell, but rather as a complex association of DNA, RNA, and proteins. This is a consequence of the role of DNA as the genetic blueprint of the cell. In order to express genetic information, the DNA is used as a template for the production of messenger RNA, which is translated by the ribosome into protein. Proteins directly involved in the process of gene expression, such as RNA polymerase and regulatory proteins, interact with DNA *in vivo* to form nu-

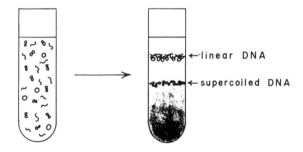

Figure 3.1
The use of cesium chloride-ethidium bromide density gradients in the separation of supercoiled plasmid DNA from linear chromosomal DNA. A DNA solution containing both linear and supercoiled DNA is mixed with cesium chloride and ethidium bromide and subjected to equilibrium centrifugation. During centrifugation, a gradient of increasing cesium concentration is established (indicated by shading). In such a gradient, DNA molecules will band at positions corresponding to their bouyant density. The binding of ethidium bromide by DNA decreases the density of the DNA and thus alters the position at which the DNA will band. This decrease in bouyant density is greater for linear and open circular DNA than for supercoiled DNA.

CHAPTER 3
DNA ISOLATION

cleo-protein complexes. DNA polymerase, DNA ligase, various unwinding and supercoiling enzymes, recombination and repair enzymes, and those proteins involved in the initiation or maintenance of DNA replication are also associated with DNA *in vivo*. In certain situations, particularly with bacteriophage and viral genomes, DNA may be encapsulated in a protein coat or tightly associated with membranes. The basic steps in the purification of DNA are: (1) the release of soluble, high molecular weight DNA from disrupted cell wall and membranes; (2) dissociation of DNA-protein complexes by denaturation or proteolysis; and (3) the separation of DNA from the other macromolecules.

Prokaryotic Chromosomal DNA

DNA of bacterial origin can be purified by different methods, depending on whether the DNA is chromosomal or extrachromosomal, such as plasmids or bacteriophage. In the purification of chromosomal DNA, the bacterial cell wall is generally weakened by freeze-thawing or by treatment with the enzyme lysozyme and the chelating agent ethylenediaminetetraacetic acid (EDTA). Cell lysis is accomplished by the addition of a detergent such as sodium dodecyl sulfate (SDS) in a buffered saline solution. Following lysis, the solution is treated with pancreatic ribonuclease to hydrolyze RNA and protease to degrade proteins. Residual proteins and oligopeptides are extracted with an organic solvent, such as phenol or an equal mixture of phenol and chloroform. Most of the protein will denature and enter the organic phase or precipitate at the interface of the organic and aqueous phases. The clear, viscous aqueous phase containing the DNA can be removed. With the addition of alcohol, the DNA will precipitate out of the aqueous phase as a white fibrous material and can be spooled out on a glass rod. Precipitation from alcohol serves to concentrate the high molecular weight DNA while removing the small oligonucleotides of DNA and RNA, detergent, and the organic solvent used in the removal of proteins. Residual detergent and salts can be removed by dialysis of the resuspended DNA solution against the desired buffer. In some instances, it may be desirable to further purify the DNA by centrifugation on isopycnic cesium chloride gradients, or hydroxylapatite chromatography.

Because of its large length-to-width ratio, long DNA molecules are extremely susceptible to shearing forces. Therefore, excessive agitation during purification will result in a reduction in the average molecular weight of the DNA molecules. As the average molecular weight of the DNA is reduced, the probability of cloning a gene intact will also decrease. The use of buffered solutions containing 10 to 25% sucrose can help to minimize the shearing of DNA as it is released from the cell. By careful handling of solutions during purification, it is possible to obtain 2 to 5 mg of DNA with an average molecular weight of 50×10^6 daltons from one liter of bacterial culture. Although this procedure can be used to isolate DNA from a variety of gram-negative bacteria, it can also be applied to gram-positive bacteria with slight modification (see Appendix 2).

Extrachromosomal DNA

During the purification of extrachromosomal elements, including plasmids and bacteriophage, it is desirable to minimize the amount of bacterial chromosomal DNA contaminating the preparation. With bacteriophage, this can often be accomplished by first purifying the phage particles from the infected bacteria, then treating the purified phage particles with protease and/or phenol to release the bacteriophage DNA. Further purification of the DNA can be accomplished by means similar to those described for chromosomal DNA. Due to its size, however, precipitated bacteriophage and plasmid DNA cannot be spooled out on a glass rod and is therefore generally recovered by centrifugation.

In the purification of plasmid DNA and bacteriophage DNA that cannot be obtained in phage particles, e.g., the replicative form of phage M13, the extrachromosomal DNA is first isolated with the bacterial chromosome and then separated. This can be accomplished by exploiting the covalently closed circular or "supercoiled" property of plasmid DNA molecules. When double-stranded DNA is exposed to high pH or high temperatures, the two strands of the helix dissociate into two single strands. This process of denaturation is reversible and under the appropriate conditions the two complementary strands of the double helix will reassociate, or reanneal, to the double-stranded form. However, the reassociation of strands is a time-dependent process, and when a denatured solution of DNA is chilled or neutralized rapidly, the complementary strands are unable to reassociate and remain in the single-strand form. When a supercoiled DNA is denatured, the two single-stranded circular DNA molecules remain intertwined. Because of the proximity of these two strands, the rate of reassociation is much faster than the reassociation of complementary linear DNA strands. Therefore, denatured supercoiled DNA can rapidly reanneal to form a double-stranded molecule. As a result of this property, a mixture of chromosomal and plasmid DNA can be denatured with alkali, then returned rapidly to a neutral pH, and the chromosomal DNA will remain predominantly in the single-stranded form. A variety of techniques, such as absorption to and elution from hydroxylapatite or selective precipitation with various detergents and salts can be used to separate the single-stranded chromosomal DNA from the double-stranded plasmid DNA.

Another property that makes supercoiled DNA different from linear DNA is its ability to bind intercalating dyes. Intercalating dyes, such as ethidium bromide and propidium iodide, are molecules that insert between the stacked purine:pyrimidine base pairs of double-stranded DNA. As each dye molecule intercalates, the overall buoyant density of the DNA molecule decreases. Supercoiled DNA will bind and accumulate dye molecules until the DNA can no longer be twisted without nicking one of the strands to relieve the tension. At this point, the molecule is saturated with the bound dye. In contrast, a linear DNA molecule is free to rotate about the axis of the helix and relieve the tension induced by the binding of an intercalating agent. Consequently, the linear molecule can bind more dye than the supercoiled DNA molecule. As a result, the density decrease caused by the binding of the dye to linear DNA will be

CHAPTER 3
DNA ISOLATION

greater than that caused by the binding of the dye to the supercoiled DNA. If linear and supercoiled DNA of the same initial buoyant density are both saturated with dye molecules, the resulting density of the linear DNA will be less than the resulting density of the supercoiled DNA.

Utilizing this principle, a mixture of supercoiled and chromosomal DNAs can be separated by equilibrium centrifugation in a cesium chloride density gradient. Because of the lower buoyant density of the dye-saturated chromosomal DNA relative to the dye-saturated plasmid DNA, the linear DNA will form a band at a different, but higher, position in the gradient than the plasmid DNA band (fig. 3.1). The DNA bands can be readily visualized by exposing the gradients to ultraviolet light. The resulting pink-orange fluorescence identifies the DNA-dye complex and allows the purified plasmid DNA to be recovered by fractionation. Since the presence of the intercalating agent will interfere with subsequent *in vitro* manipulation of the plasmid DNA, the dye must be removed. Most intercalating dyes can be removed from DNA by extraction with organic solvents such as n-butanol. Such extraction procedures are generally inefficient and require multiple extractions. A rapid and efficient means for the removal of intercalating dyes from DNA involves the use of the resin Dowex AG50 W-X8. This resin has a greater affinity for the intercalating dye than the dye has for the DNA (see Appendices 1 and 2). After removal of the dye, the cesium chloride carried over from the equilibrium gradient can be removed by dialysis and the purified DNA concentrated by ethanol precipitation.

The time-consuming step of dye-buoyant density gradient centrifugation can be circumvented by the use of a resin to which an intercalating dye has been bound. A partially purified mixture of linear and circular DNA can be bound to a resin such as acridine yellow or phenol-neutral red, and by increasing the salt concentration, linear DNA can be eluted, followed by supercoiled DNA. Again, linear and supercoiled DNA molecules can be separated on the basis of their differential binding to the intercalating agent.

At various stages in the purification of plasmid DNA it is desirable to remove contaminants such as detergents, proteins, RNA, or salts. While salts and certain small molecules can easily be removed by dialysis, this method is not generally effective for oligonucleotides and oligopeptides. As an alternative, low molecular weight contaminants can be removed by the use of gel filtration with Sephadex G-50 or Biogel A50m as the column material. In gel filtration, a matrix of beads containing small pores is used to form the column bed. When a solution of macromolecules is applied to the column and eluted with buffer, molecules too large to enter the pores in the beads flow around the beads and progress rapidly through the column. The minimum buffer volume necessary to wash such an excluded molecule through the column is the exclusion or "void" volume of the column. Molecules that are small enough to enter the pores in the beads must filter through the beads rather than flow around them. Consequently, their migration rate is considerably slower than that of the excluded molecules. When a solution containing plasmid DNA, RNA, protein, and phenol is applied to a column of Biogel A50m

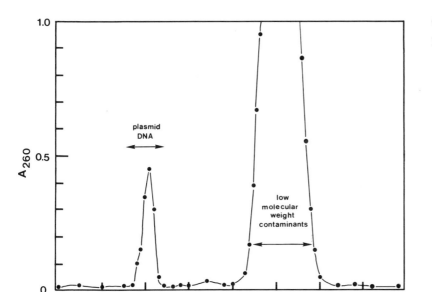

Figure 3.2
Elution profile of plasmid DNA and low molecular weight contaminants from a BioRad A50m column (20 cm × 1.0 cm). 250 μl fractions were collected at a rate of 1.6 min./ml at room temperature in an elution buffer of 50 mM Tris pH 8.0, 500 mM NaCl, 1 mM EDTA, 1 mM NaN_3. The DNA pellet from Step 1, Day 2 of Exercise 4 was resuspended in 5 ml of A50m elution buffer, isopropanol precipitated. The DNA was resuspended in 450 μl A50 Buffer and 50 μl of 80% glycerol and applied to the A50 column. The plasmid peak represents approximately 250 μg of pBR329 DNA.

agarose, molecules greater than 10^5 daltons are excluded from entry while low molecular weight contaminants ($<10^5$ daltons) enter the gel matrix. As fractions are eluted and collected, the plasmid DNA will emerge first, followed by smaller molecules, as illustrated in Fig. 3.2. Gel filtration often provides a convenient and rapid method for the removal of most contaminants from DNA solutions.

Rapid Isolation of Plasmid DNA

The procedures just described are designed primarily for the isolation of milligram quantities of very pure, plasmid DNA; however, large amounts of pure plasmid are not necessary if one simply wishes to: (1) verify the presence of a plasmid, (2) obtain an approximate estimation of the size of the plasmid, and (3) perform a limited restriction site analysis. There are several procedures that describe the isolation of small amounts of partially purified plasmid DNA from a large number of bacterial clones in a relatively short period of time. Most of these "miniscreen" procedures exploit the same principles of reversible denaturation of supercoiled DNA and selective precipitation of SDS. In most instances,

CHAPTER 3
DNA ISOLATION

contaminating chemicals and nucleic acids can be reduced to a level that allows the restriction site pattern of a particular plasmid DNA to be easily visualized by gel electrophoresis. The need for rapid plasmid isolation methods becomes apparent when large numbers of bacterial transformants have to be analyzed for the presence of recombinant DNA molecules.

Eukaryotic Chromosomal DNA

The purification of DNA from eukaryotic cells is complicated by the presence of the nucleus, chloroplasts, and/or mitochondria. While total cellular DNA can often be prepared by a modification of the detergent lysis procedure described for bacteria, it is often desirable to purify the nuclear and the organelle DNA separately. This can be accomplished in many instances by gently lysing the cell while leaving the nucleus and organelles intact. Because of the size difference, buoyant density, and sensitivity to various detergents, nuclei, chloroplasts, and mitochondria can be separately purified to homogeneity. These components can then be subjected to lysis and the DNA purified from RNA and proteins. The DNA in mitochondria and chloroplasts is generally present to some degree as covalently closed circular molecules and can be further purified from contaminating chromosomal DNA by the methods described for the purification of plasmid DNA. While these separation techniques have been applied successfully to a variety of eukaryotic cells, isolation of DNA from plant cells still remains a challenge. The presence of a rigid cell wall as found in most plants can prove to be a serious obstacle to the gentle lysis of the cell. Although no universal plant DNA extraction protocol has been devised, a number of extraction procedures are available.

Spectrophotometric Assay of DNA Concentration and Purity

Once the DNA preparation has been freed of contaminating macromolecules, the concentration of the DNA in the solution can be determined. The method most commonly used to determine DNA concentration involves the use of ultraviolet absorption spectroscopy. Just as all organic compounds have characteristic absorption spectra, the nitrogenous bases in double-stranded DNA exhibit a strong absorption maximum at a wavelength of 260 nm. At this wavelength, the extinction coefficient of DNA, $E_{260} = 20$, indicates that DNA at a concentration of 1 mg/ml will have an absorption $(A_{260}) = 20$. As the relationship between DNA concentration and A_{260} is linear to an $A_{260} = 2$, the concentration of DNA in a solution can be determined (fig. 3.3). For example, $A_{260} = 0.5$ corresponds to 25 µg/ml, $A_{260} = 0.1$ corresponds to 5 µg/ml and so on. Use of the conversion factor, 50 µg/ml = 1 A_{260} unit, enables the concentration of most DNA solutions to be determined easily. However, this relationship only applies to purified double stranded DNA with a G + C content of 50%. The presence of RNA, proteins, detergents, and organic solvents will also contribute to absorbance at this wavelength. Since the

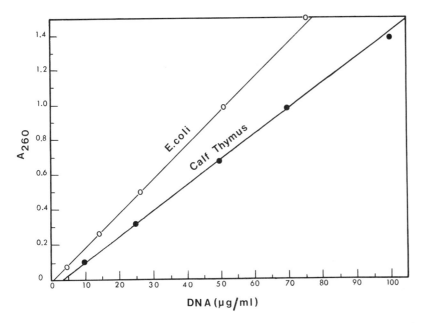

Figure 3.3
Standard absorbance versus concentration curves for *E. coli* DNA (guanine + cytosine = 51%) and calf thymus DNA (guanine + cytosine = 43.6%). Serial dilution of the DNAs were made in $KHPO_4$ buffer pH 7.0, 150 mM NaCl.

absorption maxima for DNA and protein are 260 nm and 280 nm, respectively, an approximate measure of the purity of the isolated DNA can be obtained by determining the A_{260}/A_{280} ratio. Pure *E. coli* DNA has $A_{260}/A_{280} = 1.95$. This ratio, however, is dependent on the overall base composition of the DNA and will vary with different organisms.

References

Birnboim, H. C., and Doly, J. 1979. A rapid alkaline extraction procedure for recombinant plasmid DNA. Nucl. Acids Res. 7:1513–1525.

Cryer, D. R., Eccleshall, R., and Marmur, J. 1975. Isolation of yeast DNA. Methods Cell Biol. 12:39–44.

Currier, T. C., and Nester, E. W. 1976. Isolation of covalently closed circular DNA of high molecular weight from bacteria. Anal. Biochem. 76:431–441.

Guerry, P., LeBlanc, D. J., and Falkow, S. 1973. General method for the isolation of plasmid deoxyribonucleic acid. J. Bacteriol. 116:1064–1066.

Hansen, J., and Olsen, R. 1978. Isolation of large bacterial plasmids and characterization of the P2 incompatibility group plasmids pMG1 and pMG5. J. Bacteriol. 135:227–238.

Holmes, D. S., and Quigley, M. 1981. A rapid boiling method for the preparation of bacterial plasmids. Anal. Biochem. 114:193–197.

Kado, C. I., and Liu, S.-T. 1981. Rapid procedure for the detection and isolation of large and small plasmids. J. Bacteriol. 145:1365–1373.

Kislev, N., and Rubenstein, I. 1980. Utility of ethidium bromide in extraction from whole plants of high molecular weight maize DNA. Plant Physiol. 66:1140–1143.

CHAPTER 3

DNA ISOLATION

Manning, J. E., Wolstenholme, D. R., Ryan, R. S., Hunter, J. A., and Richards, O. C. 1971. Circular chloroplast DNA from *Euglena gracilis*. Proc. Natl. Acad. Sci. USA 68:1169–1173.

Marmur, J. 1961. A procedure for the isolation of deoxyribonucleic acid from microorganisms. J. Mol. Biol. 3:208–218.

McMaster, G. K., Samulski, R. J., Stein, J. L., and Stein, G. S. 1980. Rapid purification of covalently closed circular DNAs of bacterial plasmid and animal tumor viruses. Anal. Biochem. 109:47–54.

Murray, M. G., and Thompson, W. F. 1980. Rapid isolation of high molecular weight plant DNA. Nucl. Acids Res. 8:4321–4325.

Smith, C. A., Jordan, J. M., and Vinograd, J. 1971. *In vivo* effects of intercalating drugs on the superhelix density of mitochondrial DNA isolated from human and mouse cells in culture. J. Mol. Biol. 59:255–272.

Thomas, C. A., and Abelson, J. 1966. The isolation and characterization of DNA from bacteriophage. In Procedures in Nucleic Acid Research, G. L. Cantoni and D. R. Davies, eds., pp. 553–576. Harper and Row, New York.

Thomas, C. A., Bernes, K. I., and Kelley, T. J. 1966. Isolation of high molecular weight DNA from bacteria and cell nuclei. In Procedures in Nucleic Acid Research, G. L. Cantoni and D. R. Davies, eds., pp. 535–540. Harper and Row, New York.

Vincent, W. S., and Goldstein, E. S. 1981. Rapid preparation of covalently closed circular DNA by acridine yellow affinity chromatography. Anal. Biochem. 110:123–127.

Yamamoto, K. R., Alberts, B. M., Benzinger, R., Lawhorne, L., and Treiber, G. 1970. Concentration and purification of vesicular stomatitis virus by polyethylene glycol "precipitation." Virol. 40:734–744.

EXERCISE 2

Isolation and Purification of E. coli *Chromosomal DNA*

Materials

Luria Broth (LB):   80 ml.

SET Buffer: 20% sucrose, 50 mM Tris-HCl pH 7.6, 50 mM EDTA.

RNase Buffer: Pancreatic ribonuclease A, 10 mg/ml, in 0.1 M Na acetate, 0.3 mM EDTA, pH 4.8, preheated to 80°C for 10 minutes.

TEN Buffer: 10 mM Tris-HCl pH 7.6, 1 mM EDTA, 10 mM NaCl.

Chloroform/n-amyl Alcohol: 24 to 1 mixture, respectively.

Lysozyme: 5 mg/ml in TEN Buffer.

Pronase: 2 mg/ml in TEN buffer preheated to 37°C for 15 minutes.

Sodium Dodecyl Sulfate (SDS):   25%.

NaCl: 5 M.

Ethanol: 95%, stored in a freezer.

Water (ddH$_2$O):   Sterile, double distilled.

Dry Ice.

Note: Many of these reagents and materials will be used for subsequent exercises.

Protocol

DAY 1

1. Grow *E. coli*-AB257 in an 80 ml culture of LB to late log phase of cell growth (approximately 5–8 × 10^8 cells/ml).

2. Using 2, 50 ml polypropylene centrifuge tubes, harvest the cells by centrifugation at 3,000 rpm (3 krpm) for 10 minutes (or 10 krpm for 5 minutes).

3. Resuspend each cell pellet in 10 ml of SET Buffer by gentle vortexing. Combine the two cell suspensions into 1 centrifuge tube and harvest the cells as previously described.

4. Freeze the cell pellet in a dry ice-ethanol bath for 2 minutes. Thaw the pellet in warm water, resuspend the cells in 2 ml of SET Buffer, and set on ice.

EXERCISE 2

ISOLATION AND PURIFICATION OF *E. COLI* CHROMOSOMAL DNA

5. Add 0.2 ml of lysozyme and 0.1 ml of RNase Buffer and incubate on ice for 15 minutes. Pour the cell suspension into a 25 ml screw-cap Corex tube and add 0.05 ml of SDS. Incubate the mixture at 37°C with gentle shaking for 3 to 6 hours.

6. Add 0.3 ml of pronase and 1.5 ml of chloroform/n-amyl alcohol and continue the 37°C incubation (with gentle shaking) overnight (10 to 16 hours).

DAY 2

1. Add 1 ml of ddH$_2$O and 2 volumes (10 ml) of chloroform/n-amyl alcohol. Tighten the cap and invert the tube gently for 5 minutes. Centrifuge the tube (in an appropriate rubber adapter) at 9 krpm for 5 minutes to separate the phases. Remove the aqueous (upper) phase with a wide-mouth pipet and place in a clean 25 ml screw-cap Corex tube. Repeat the chloroform extraction two more times.

2. Again with a wide-mouth pipet (to prevent shearing), remove the aqueous phase and place it into a 30 ml Corex tube. Add 0.2 ml of NaCl and 2 volumes (approximately 11 to 12 ml) of ice-cold 95% ethanol. Mix the solution gently but thoroughly and let it stand on ice for 5 minutes.

3. With a glass rod (or flame-bent end of a disposable pasteur pipet) spool out the fibrous strands of precipitated DNA. Rinse the DNA free of low molecular weight contaminants by dipping the DNA into a tube of fresh 95% ethanol.

4. Dissolve the precipitated DNA in 3 ml of TEN Buffer and store it at 4°C over 0.1 ml of chloroform. Pure DNA will dissolve in the TEN Buffer within 30 minutes. DNA contaminated with protein may take several hours to go into solution.

Note: Procedures for isolating high molecular weight DNA from *Bacillus subtilis, Drosophila melanogaster,* and *Saccharomyces cerevisiae* can be found in Appendix 2.

EXERCISE 3

Determination of the Concentration and Purity of DNA by Ultraviolet Absorption Spectroscopy

Materials

DNA: In TEN Buffer.

Potassium Phosphate (KP) Buffer: 10 mM K_2HPO_4/KH_2PO_4 pH 7.0, 150 mM NaCl.

Protocol

DAY 1

1. Take 0.1 ml of purified bacterial chromosomal DNA and add it to a tube containing 1.9 ml of KP Buffer. Mix thoroughly.

2. Use 1 ml of this solution to determine the ultraviolet absorbance at 260 nm and 280 nm of wavelength.

3. Using the relationship 50 µg DNA = 1 absorbance unit at A_{260}, calculate the DNA concentration/ml, the total yield of DNA, and the 260 nm/280 nm ratio.

EXERCISE 4

Large-Scale Plasmid Purification

Materials

SET Buffer: 20% sucrose, 50 mM Tris-HCl pH 7.6, 50 mM EDTA.

Lytic Mixture: 1% SDS, 0.2 N NaOH (make fresh mixture every two weeks).

K (potassium) Acetate Buffer: 3.0 M, pH 4.8.

RNase Buffer: 10 mg/ml in 0.1 M Na acetate, 0.3 mM EDTA, pH 4.8, preheat to 80°C for 10 minutes.

TEN Buffer: 10 mM Tris-HCl pH 7.6, 1 mM EDTA, 10 mM NaCl.

Chloroform/Buffer Saturated Phenol: 1 to 1 mixture.

Isopropanol.

Ether: Hydrated.

Ethanol: 70%.

Luria Broth (LB): 200 ml.

NaCl: 5 M.

Ice.

Protocol

DAY 1

1. Grow a 200 ml LB culture of RR1/pBR329 to stationary phase of cell growth. (Chloramphenicol amplification of plasmid DNA is not desirable.)

2. Harvest cells by centrifugation in 250 ml polypropylene centrifuge bottles at 6.5 krpm, for 5 minutes.

3. Thoroughly resuspend the cell pellet with 30 ml of SET Buffer. Add 65 ml of Lytic Mixture.

4. Vortex the solution thoroughly and incubate at 50°C for 30 minutes, swirling occasionally.

5. Transfer the solution to an ice bath and let it stand for 25 minutes.

6. Neutralize the solution by the slow addition of 10–20 ml of 3.0 M K Acetate Buffer, swirling the solution constantly. The SDS and denatured chromosomal DNA will precipitate out of the solution at this stage. Monitor the neutralization step with pH indicator paper.

7. Let the solution stand in the ice bath for 40 minutes, with occasional swirling to even the solution temperature.
8. Spin out the precipitate by centrifugation at 4°C, 6.5 krpm for 20 minutes. Carefully pour the supernatant through a Kimwipe and funnel into a 250 ml centrifuge bottle.
9. Add 0.2 ml of RNase Buffer and incubate at 40°C for 20 minutes. Extract residual proteins by adding 50 ml of a 1:1 mixture of chloroform and buffer saturated phenol.
10. Cap the bottle tightly and swirl the mixture vigorously for 2 minutes. (*Caution:* Use gloves to avoid contact with phenol.) Separate the aqueous and organic phases by centrifugation at 6.5 krpm at 4°C for 10 minutes.
11. Remove and measure the aqueous (upper) phase with a 25 ml pipet. Transfer the aqueous phase to another 250 ml bottle. Store at 4°C.

Note: All subsequent manipulations should be carried out at room temperature unless stated otherwise.

EXERCISE 4

LARGE-SCALE PLASMID PURIFICATION

DAY 2

1. Add an equal volume of isopropanol (room temperature) to the aqueous phase (room temperature) in a 250 ml bottle. Mix well and let it stand for 5 minutes before centrifuging at 6.5 krpm for 20 minutes at 20°C.
2. Pour off the supernatant and resuspend the plasmid DNA pellet in 8 ml of TEN Buffer. Add 0.3 ml of 5 M NaCl and transfer to a 30 ml Corex tube.
3. Add 8 ml of hydrated ether and vortex for about 30 seconds. Pipet off the ether (upper) phase. (*Caution:* Avoid open flames when using ether.)
4. Blow off residual ether and add an equal volume of isopropanol. Mix and let it stand for 5 minutes. Centrifuge for 20 minutes at 6.5 krpm at 20°C. Add another 8 ml of TEN Buffer to the pellet and precipitate once again with an equal volume of isopropanol.
5. Wash the plasmid DNA pellet with 10 ml of 70% ethanol (room temperature). Centrifuge 5 minutes at 6.5 krpm at 20°C. Pour off ethanol and resuspend the pellet in 1 ml of TEN Buffer. The DNA should be dialyzed against 2 liters of TEN Buffer at room temperature for at least 3 hours.

Note: This exercise should be repeated with the strains RR1/pPV33 and RR1/pPV501.

EXERCISE 5

Rapid Isolation of Plasmid DNA (Miniscreen)

Materials

SET Buffer: 20% sucrose, 50 mM Tris-HCl pH 7.6, 50 mM EDTA.

Lytic Mixture: 1% SDS, 0.2 N NaOH (make fresh mixture every two weeks).

Na (sodium) Acetate Buffer: 3.0 M, pH 4.8.

RNase Buffer: Pancreatic ribonuclease A, 10 mg/ml in 0.1 M Na acetate, 0.3 mM EDTA, ph 4.8, preheated to 80°C for 10 minutes.

Isopropanol.

Ethanol: 70%.

Water (ddH$_2$O): Sterile double distilled.

Luria Broth (LB): 3 ml.

Ice.

Protocol

DAY 1

1. Grow to stationary phase of cell growth, three 1.5 ml LB cultures of RRl containing the plasmids pBR329, pPV33, and pPV501, respectively.
2. Transfer each culture to separate microcentrifuge (microfuge) tubes and spin 0.5 minute. (All centrifugations will be done in 1.5 ml microfuge tubes.)
3. Resuspend the cell pellet in 1 ml SET Buffer by thoroughly vortexing for 1 minute.
4. Centrifuge the cells for 1 minute and resuspend in 150 µl SET Buffer.
5. Add 5 µl RNase buffer and vortex. Add 350 µl of Lytic Mixture at room temperature. Vortex briefly. Solution will become clear.
6. Place on ice bath and incubate 10 minutes.
7. Add 250 µl Na-Acetate Buffer and invert the tube several times. Incubate 30 minutes on ice. SDS and denatured chromosomal DNA will precipitate out of solution at this stage.
8. Centrifuge the solution 5 minutes at 4°C. Pipet the supernatant to a clean microfuge tube (approximately 700 µl).

9. Add an equal volume of isopropanol. Invert tubes several times and centrifuge 5 more minutes at room temperature. Invert tubes and drain to remove isopropanol.

10. Wash the DNA pellet by adding 1 ml 70% ethanol. Centrifuge 5 minutes at room temperature. Decant the ethanol and vacuum dry the tubes for 10 minutes. Resuspend in 20 µl of ddH$_2$O.

11. Use 1–2 µl of the "miniscreen" DNA for each restriction digest.

    Note: A second wash with 70% ethanol may be required if restriction digests do not go to completion.

    This method can also be used to isolate M13 RF DNA.

EXERCISE 5

RAPID ISOLATION OF PLASMID DNA (MINISCREEN)

CHAPTER 4

Restriction Endonucleases

The class of enzymes known as restriction endonucleases has played a key role in the development of recombinant DNA technology. These enzymes have been found in nearly every microorganism examined and are known to catalyze double-strand breaks in DNA, to yield restriction fragments. Because a restriction enzyme cleavage pattern is specific for a given DNA and enzyme, a restriction fragment, when isolated on a preparative scale, represents a *homogeneous* population of DNA molecules. Whole genomes can now be subdivided into smaller, discrete pieces so that genes can be isolated, cloned, and characterized by nucleotide sequence analysis. Therefore, the discovery and application of restriction enzymes represents the single most important factor that has helped to revolutionize molecular biology and create the "new genetics."

Host Restriction/Modification

The first observations suggesting the existence of restriction enzymes were made in the early 1960s by W. Arber and collaborators while studying the efficiency of plating of the bacteriophage lambda on different strains of *Escherichia coli*. These observations, summarized in Fig. 4.1, indicated that phage grown on a particular strain of bacteria were able to infect that strain more efficiently than a different strain. Subsequent work revealed that this reduced efficiency of plating on different strains of bacteria was due to the existence of restriction/modification systems.

The restriction/modification system of *E. coli* involves the action of three linked genes, *hsd*R, *hsd*M, and *hsd*S, located at 98 minutes on the genetic map of *E. coli*. The product of *hsd*M methylates DNA in a specific manner to protect it from cleavage by the endonuclease encoded by *hsd*R. The product of *hsd*S possesses neither restriction nor modification activity but is required by these two enzymes for site recognition. DNA entering the cell by means of infection, conjugation, or transformation

CHAPTER 4
RESTRICTION ENDONUCLEASES

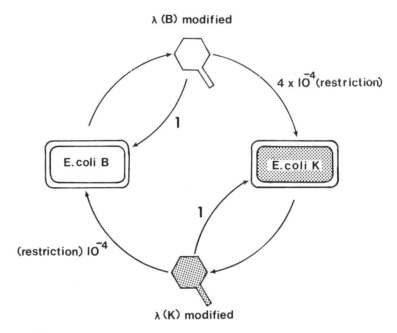

Figure 4.1
Diagrammatic representation of the host restriction/modification systems of *E. coli* K and B. Numbers indicate the degree of restriction in terms of how well the modified phage form plaques (efficiency of plating) on the different hosts.

will be cleaved and rapidly degraded if it is not modified in the proper pattern. Genetic analysis of the restriction/modification system of *E. coli* K-12 has revealed the following phenotypes:

$r_k^+ m_k^+$ = wild type: The cell possesses functional restriction and modification activities.

$r_k^- m_k^+$ = restriction deficient mutant: Only the modification system is functional. Foreign DNA introduced into these cells will be modified in the K-12 pattern and will subsequently be resistant to the K-12 restriction enzyme. Mutant strains of this type are frequently used in transformation experiments to minimize degradation of entering DNA.

$r_k^- m_k^-$ = restriction and modification deficient mutant: Such strains cannot degrade or modify foreign DNA and are also commonly used in transformation experiments.

$r_k^+ m_k^-$ = restriction proficient but modification deficient: This gene configuration, also known as the "suicide genotype," can only be obtained using conditional lethal mutations such as temperature sensitivity.

The precise role of restriction enzymes in the cell has not yet been fully elucidated. It is clear from the work of Arber that these enzymes may serve to protect a cell from infection by bacteriophage. It has been suggested that these enzymes may also play a role in promoting "site-specific illegitimate recombination," allowing incoming DNA to be cleaved and incorporated into the chromosome. This form of genetic exchange would circumvent the barriers that normally prevent the exchange of genetic information between unrelated organisms.

CHAPTER 4

RESTRICTION ENDONUCLEASES

The endonucleases involved in the host restriction/modification system have been designated Type I endonucleases. These enzymes are large multimeric proteins that are capable of both cleaving and modifying DNA. Enzymes of this type require ATP, Mg^{++}, and the co-factor S-adenosylmethionine (SAM). In 1968, H. O. Smith and co-workers discovered another type of restriction endonuclease in the bacterium *Haemophilus influenzae* type d. This enzyme was designated Type II because it was a smaller monomeric protein that required only Mg^{++} for activity. Unlike the Type I enzyme, this class was capable of cleaving DNA at specific sites and was shown to produce DNA fragments with cohesive termini. Because of the site-specific nature of the cleavage, Type II restriction enzymes have become one of the primary tools in the analysis and restructuring of DNA.

Type II Restriction Endonucleases

There are approximately 250 restriction enzymes that have been identified to date. In order to simplify the naming of these enzymes, a nomenclature has been developed that is based on an abbreviation of the name of the organism from which the enzyme was isolated. The first initial of the genus and the first two initials of the species form the basic name. When the enzyme is present in a specific strain, these three italicized letters may be followed by a strain designation. The third portion of the name is reserved for a Roman numeral indicating the order of discovery of the enzyme in the particular strain. For example, *Hae*II is purified from the strain *Haemophilus aegypticus* and is one of three restriction enzymes present in the strain. *Hin*fI is the enzyme purified from *Haemophilus influenzae* strain f. Perfect isoschizomers are enzymes that have been purified from different organisms but still recognize and cleave at the same sequence in the DNA. Enzymes such as *Hin*dIII and *Hsu*I cleave at the same position in the recognition sequence, in which case the resulting termini are indistinguishable. Imperfect isoschizomers may recognize the same sequence but cleave at different nucleotides within the sequence, as is the case with *Xma*I and *Sma*I (Table 4.1). Two restriction enzymes may recognize the same sequence, but cleave in a manner that is dependent on the methylation of certain bases within the recognition sequence. For example, *Msp*I will cleave at the sequence CCGG, but *Hpa*II will cleave this sequence only if the internal C has not been methylated by cytosine methylase. Table 4.1 lists the recognition sites and cleavage points for several commonly used restriction enzymes. A more complete list can be found in Appendix 4.

The recognition sites for Type II restriction endonucleases are generally 4 (tetramer), 5 (pentamer), or 6 (hexamer) base pairs in length, and most share the feature of two-fold rotational (dyad) symmetry. A region of DNA is said to have dyad symmetry when the nucleotide sequence of one DNA strand is the same as the nucleotide sequence of its complementary strand when each strand is read in the 5' to 3' direction.

CHAPTER 4
RESTRICTION ENDONUCLEASES

Table 4.1

Specific endonucleases and their recognition sequences

TETRANUCLEOTIDE		PENTANUCLEOTIDE		HEXANUCLEOTIDE	
*Alu*I	AG↓CT	*Eco*RII	↓CC(A/T)GG	*Ava*I	C↓PyCGPuG
*Hae*III	GG↓CC	*Hinf*I	G↓ANTC	*Bam*HI	G↓GATCC
*Hha*I	GCG↓C			*Bgl*II	A↓GATCT
*Hpa*II	C↓CGG			*Bal*I	TGG↓CCA
*Mbo*I	↓GATC			*Eco*RI	G↓AATTC
*Taq*I	T↓CGA	*Hph*I	GGTGA(N)$_8$↓	*Hind*III	A↓AGCTT
		*Mbo*II	GAAGA(N)↓$_8$	*Hpa*I	GTT↓AAC
				*Pst*I	CTGCA↓G
				*Xma*I	C↓CCGGG
*Dpn*I	GA↓TC			*Sma*I	CCC↓GGG
	when			*Hae*II	PuGCGC↓Py
	modified			*Hinc*II	GTPy↓PuAC

For example, the recognition site for *Eco*RI has true dyad symmetry,

$$\begin{array}{ll} 5' \ G\overset{\downarrow}{A}\overset{*}{A} \ | \ TTC \ 3' \\ 3' \ CTT \ | \ \underset{*}{A}\underset{\uparrow}{A}G \ 5' \end{array}$$

with the axis of rotation located between the central AT base pairs. (*Eco*RI cleavage and methylation sites are indicated by arrows and asterisks, respectively.) In the case of some penta- and hexanucleotide recognition sites, there may be internal (*Eco*RII, *Hinc*II) or external (*Hae*II) degeneracy of the dyad symmetry. While this degeneracy does not affect the action of the endonuclease and methylase, it does increase the frequency at which these sites will occur in DNA.

The frequency of occurence of a restriction cleavage site in DNA can be estimated using the formula:

Site frequency = $\frac{1}{4}^N$

where

N = the length of the restriction site sequence

For a tetrameric sequence the frequency is $\frac{1}{4}^4$ bases, or an average of 1 site for every 256 base pairs, and for a hexameric sequence the frequency is $\frac{1}{4}^6$ bases, or approximately 1 site every 4,096 base pairs. Using this principle, it is possible to estimate the number of cleavage sites expected in a molecule of known size:

Size of molecule (in base pairs) × Frequency of site = Number of sites.

For example, there should be approximately 1 *Eco*RI site in a 5,000 base pair fragment, 2 in a 10,000 base pair fragment, and 5 in a 20,000 base pair fragment.

CHAPTER 4

RESTRICTION ENDONUCLEASES

Unfortunately, empirical observation often contradicts mathematical theory, particularly when the theory is based on the random incorporation of nucleotides during DNA synthesis. The 40 kilobase pair (kb) bacteriophage T7 should have about 10 cleavage sites for each of the enzymes *Eco*RI, *Hin*dIII, and *Bam*HI, but in fact, it has no sites for these enzymes. Tetrameric recognition sites, however, appear with the expected frequency. It is also apparent that restriction sites are not distributed randomly in DNA, but tend to cluster in and around structural genes. In the 36 kb R factor pSa, the hexameric recognition sites are clustered in a 10 kb region that encodes the four drug resistance genes of the plasmid. The remaining 26 kb of the plasmid is involved in plasmid replication and transfer and contains few hexamer-specific cleavage sites. Finally, the frequency of restriction sites containing the CG dinucleotide (*Hha*I, *Hpa*II, and *Hae*II) is much lower in eukaryotes than prokaryotes. This can be explained by the fact that eukaryotes, particularly mammals, avoid codons containing the CG dinucleotide.

While restriction endonucleases are known to make double-strand scissions in DNA, these cuts do not all occur in the same fashion. In general, there are three positions for cleavage within a recognition sequence: (1) at the center or axis of the sequence to give a blunt-end cut, such as that made by *Hae*III; (2) to the left of the center to give cohesive termini with a protruding 5' phosphate residue, such as with *Eco*RI; or (3) to the right of the center to give cohesive termini with an exposed 3' hydroxyl residue, such as with *Pst*I.

```
     (1)                    (2)                     (3)
        ↓                      ↓                       ↓
-GG3' 5'CC-         -G3'  5'AATTC-          -CTGCA3' 5'G-
-CC5' 3'GG-         -CTTAA5' 3'G-           -G5' 3'ACGTC-
     ↑                      ↑                       ↑
```

Some restriction enzymes, such as *Mbo*II, have unusual recognition and cleavage patterns. *Mbo*II recognizes the sequence GAAGA (note the lack of dyad symmetry) and cleaves between the 8th and 9th residues to the right, leaving a one base 3' extension:

```
5'-GAAGANNNNNNNN↓-3'
3'-CTTCTNNNNNNN↑-5'
```

The different cleavage points within recognition sequences can play an important role in the manipulation of DNA fragments. For example, a protruding 5' phosphate is more readily labelled with γ-^{32}P ATP by DNA kinase than is a protruding 3' OH, whereas the protruding 3' OH is the preferred substrate for the N-terminal transferase used to attach polynucleotide tails to a DNA fragment. In a similar manner, *E. coli* DNA polymerase I can be used to convert a cohesive end to a blunt end only when a recessed 3' OH residue is provided. Since the initiation of DNA synthesis requires a 3' OH template (base paired) primer, a protruding 3' OH, such as that produced by *Pst*I cleavage, cannot be used by DNA polymerase.

It should be emphasized that the published recognition sequences for Type II restriction endonucleases were determined using specific digestion conditions. Sometimes, when these conditions are changed, the specificity of the enzyme may be reduced so that additional sites can be recognized and cleaved. The *Eco*RI endonuclease normally recognizes the sequence GAATTC, but under conditions of low salt (<50 mM), high pH (>8), and with glycerol present, a degenerate enzyme activity, *Eco*RI*, can be produced. The *Eco*RI* activity recognizes sites that differ in a single position from the original sequence. Any substitution can occur at any one of the six positions, with the exception of T→A or A→T changes in the central tetramer. Since *Eco*RI* sites are degenerate hexamers of the *Eco*RI site (e.g., GAATTA, AAATTC, and GAGTTC), the occurrence of potential cleavage sites is 15 times more frequent than *Eco*RI sites. In addition to *Eco*RI*, a similar degenerate activity (*Bam*HI*) has been reported for the enzyme *Bam*HI. Certain enzymes are optimally active only under relatively specific reaction conditions, and alteration of pH or ionic conditions may drastically affect the activity of the enzyme. In order to obtain the expected digestion specificity with the optimal reaction rate, it is important to adhere to the reaction conditions recommended for a particular restriction enzyme.

Purification of Restriction Endonucleases

Like most of the enzymes used in the *in vitro* manipulation of DNA, restriction enzymes are generally purchased from one of a variety of commercial suppliers. Because of the high affinity of these proteins for DNA and DNA analogs, the preparation of Type II restriction endonucleases is a relatively simple procedure. Purification generally involves lysis of the cells, followed by removal of the debris by high-speed centrifugation. A substantial purification results when the DNA-binding proteins in the lysate are bound to DNA-cellulose or phosphocellulose under conditions of low salt concentration and subsequently eluted with a gradient of increasing salt concentration. Aliquots of the eluted fractions are incubated with a plasmid or phage DNA, and then examined by gel electrophoresis for the presence of endonuclease activity (fig. 4.2). The active fractions generally contain contaminating exo- and endonucleases at this stage, so they are pooled and subjected to further purification. Subsequent purification steps may involve ammonium sulphate fractionation or chromatography with hydroxylapatite, DEAE-sephadex, or other resins. In a typical purification, 5,000 to 200,000 units of enzyme might be prepared in 3 to 4 days from an initial batch of 50 grams of bacteria. The resulting enzyme is generally dialyzed into a storage buffer containing 50% glycerol and stored at $-20°C$. The purified protein is most stable in a concentrated form (10 to 100 units/μl) and may be diluted to 1 to 3 units/μl prior to use to avoid using excessive amounts of enzyme during digestion of DNA samples. Reaction buffers for various enzymes can be found in Appendix 4.

Terminating Restriction Endonuclease Reactions

Following digestion of a DNA sample, it is often desirable to inactivate the enzyme prior to further manipulation of the DNA. Most restriction enzymes can be inactivated by incubation at 65°C for 5 minutes. Certain enzymes, such as *Bam*HI and *Hae*III, are not readily heat-inactivated and

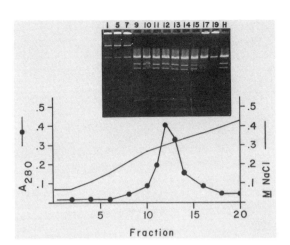

Figure 4.2
Assay for the restriction endonuclease *Hin*dIII. A partially purified preparation of *Hin*dIII enzyme was bound to a column containing Biorex 70 resin. Protein was eluted from the column with a linear gradient of 0.06 to 0.50 M NaCl. The protein elution profile was determined, and 3 µl samples of fractions were assayed for *Hin*dIII activity by incubation with 0.2 µg of plasmid DNA. After a 20-minute incubation at 37°C, the DNA samples were examined by gel electrophoresis, shown above the elution profile of the column. *Hin*dIII activity was present in fractions 9–17. The most active fractions (10–14) were pooled and dialyzed against *Hin*dIII storage buffer. H = *Hin*dIII-treated control.

CHAPTER 4
RESTRICTION ENDONUCLEASES

must be inhibited by other means. The Stop Mix added to DNA samples prior to gel electrophoresis (see Chapter 5) generally contains a protein denaturing agent such as SDS or urea to inactivate the enzyme. Although these agents are effective means of inactivating restriction enzymes, they can be difficult to remove from the DNA sample if it is necessary to perform further manipulations with the DNA. When it is necessary to inactivate an enzyme without affecting subsequent reactions, a DNA sample can be extracted with an equal volume of phenol/chloroform to remove protein. Residual phenol, which will inhibit further enzymatic reactions, can be removed by extraction of the sample with ether, and the DNA sample precipitated with cold ethanol or isopropanol in the presence of 0.2 M NaCl. Following centrifugation, the DNA pellet can be washed once with 70% ethanol to remove residual salt, then dried and resuspended in buffer. The sample is now free of protein and ready for the next enzyme treatment.

Generating New Restriction Enzyme Cleavage Sites

In the early days of genetic engineering, experiments were designed according to the availability of a small number of restriction enzymes. If the DNA molecule of interest did not possess cleavage sites for these enzymes, one had to resort to more complicated means of cloning, such as AT and GC tailing. This is no longer a serious problem with the more than 40 different restriction enzymes now commercially available. However, some experiments require the conversion of one particular restric-

Table 4.2
Regeneration of a hexamer cleavage site using DNA polymerase I-Klenow fragment and blunt-end ligation

	HEXAMER SITE TO BE REGENERATED			DONOR OF MISSING BASE PAIR			
ENZYME	TERMINUS	AFTER DNA POL I	MISSING BASE PAIR	TERMINUS	AFTER DNA POL I	ENZYME	LIGATED PRODUCT
EcoRI	G CTTAA	GAATT CTTAA	C G	CT GA	—	AluI	GAATTCT CTTAAGA
				CG GC	—	ThaI	GAATTCC CTTAAGC
				CCGGG C	CCGGG GGCCC	XmaI	GAATTCCGGG CTTAAGGCCC
				CTAGA T	CTAGA GATCT	XbaI	GAATTCTAGA CTTAAGATCT

tion site to another. Frequently, this involves the conversion of blunt-cut sites to staggered-cut sites in order to confer cohesive-ends to a previously blunt-ended restriction fragment. This manipulation can be performed using: (1) synthetic oligonucleotides called restriction site "linkers"; or (2) DNA modifying enzymes.

Restriction site linkers in the form of synthetic octamers and decamers are available for many of the common restriction endonucleases. However, restriction site conversion can be performed without linkers by using restriction enzymes in combination with common DNA modifying enzymes such as DNA polymerase I-Klenow fragment or T4 DNA polymerase (see Appendix 4 for a list of additional DNA modifying enzymes). As shown in Table 4.2, the cohesive terminus generated by EcoRI digestion can be made blunt by DNA synthesis primed from the recessed 3' OH (Klenow fragment and T4 DNA polymerase are used in these reactions because they do not form artifactual hair-pin structures like native DNA polymerase I from E. coli). When this filled-in terminus is ligated to a blunt-end fragment that provides the missing base of the recognition site (i.e., 5' cytidylic acid), the EcoRI site is regenerated and the AluI site converted to an EcoRI site simultaneously. This enzymatic manipulation can also be used to convert the 5'C donors listed in Table 4.2 to staggered-cut sites such as AvaII, BamHI, BstEII, HinfI and XhoII. The use of restriction enzymes in conjunction with other DNA modifying enzymes can allow extensive *in vitro* restructuring of DNA molecules.

References

Arber, W. 1979. Promotion and limitation of genetic exchange. Science 205:361–365.

Boyer, H. W. 1971. DNA restriction and modification mechanisms in bacteria. Ann. Rev. Micro. 25:153–176.

Chang, S., and Cohen, S. N. 1977. *In vivo* site specific genetic recombination promoted by the *Eco*RI restriction endonuclease. Proc. Natl. Acad. Sci. USA 74:4811–4815.

Gardner, R. C., Howarth, A. J., Messing, J., and Shepherd, R. J. 1982. Cloning and sequencing of restriction fragments generated by *Eco*RI* DNA 1:109–115.

Greene, P. J., Gupta, M., Boyer, H. W., Brown, W. E., and Rosenberg, J. M. 1981. Sequence analysis of the DNA encoding the *Eco*RI endonuclease and methylase. J. Biol. Chem. 256:2143–2153.

Greene, P. J., Heyneker, H. L., Bolivar, F., Rodriguez, R. L., Betlach, M. C., Covarrubias, A. A., Backman, K., Russell, D. J., Tait, R., and Boyer, H. W. 1978. A general method for purification of restriction enzymes. Nucl. Acids Res. 5:2373–2380.

Hedgpeth, J., Goodman, H. M., and Boyer, H. W. 1972. DNA nucleotide sequence restricted by the RI endonuclease. Proc. Natl. Acad. Sci. 69:3448–3452.

Kelly, T. J., and Smith, H. O. 1970. A restriction enzyme from *Haemophilus influenzae*. II. Base sequence of the recognition site. J. Mol. Biol. 51:393–409.

Meselson, M., and Yuan, H. 1968. DNA restriction enzyme from *E. coli*. Nature 217:1110–1114.

Nathans, D. 1979. Restriction endonucleases, simian virus 40, and the new genetics. Science 206:903–909.

Osterlund, M., Luthman, H., Nilsson, S. V., and Magnusson, G. 1982. Ethidium-bromide-inhibited restriction endonucleases cleave one strand of circular DNA. Gene 20:121–125.

Polisky, B., Greene, P., Garfin, D. E., McCarthy, B. J., Goodman, H. M., and Boyer, H. W. 1975. Specificity of substrate recognition by the *Eco*RI restriction endonuclease. Proc. Natl. Acad. Sci. USA 72:3310–3314.

Roberts, R. J. 1979. Directory of restriction endonucleases. In Methods in Enzymology, R. Wu, ed., 68:27–41.

Roberts, R. J. 1980. Restriction and modification enzymes and their recognition sequences. Gene 8:329–343.

Smith, H. O. 1979. Nucleotide sequence specificity of restriction endonucleases. Science 205:455–462.

Smith, H. O., and Nathans, D. 1973. A suggested nomenclature for bacterial host modification and restriction systems and their enzymes. J. Mol. Biol. 81:419–423.

Wartell, R. M., and Reznikoff, W. S. 1980. Cloning DNA restriction endonuclease fragments with protruding single stranded ends. Gene 9:309–319.

EXERCISE 6

Restriction Endonuclease Digestion of Plasmid and E. coli Chromosomal DNA

Materials

*Pst*I Reaction Buffer (10X): 200 mM Tris-HCl pH 7.5, 100 mM $MgCl_2$, 500 mM $(NH_4)_2 SO_4$.

*Hin*dIII Reaction Buffer (10X): 200 mM Tris-HCl pH 7.5, 70 mM $MgCl_2$, 500 mM NaCl, 70 mM 2-mercaptoethanol.

*Pst*I Endonuclease: Diluted to 1 unit/μl.

*Hin*dIII Endonuclease: Diluted to 1 unit/μl.

Dilution Buffer: 10 mM Tris-HCl pH 7.5, 100 mM NaCl, 0.1 mM Na_2EDTA, 1 mM dithiothreitol (Cleland's reagent), 50% glycerol, 500 μg/ml BSA.

Reaction Stop Mix: 5M urea, 10% glycerol, 0.5% SDS, 0.025% xylene cyanole FF, 0.025% bromphenol blue WS.

Miniscreen Plasmid DNA: Isolated in Exercise 5.

Purified Plasmid DNA: Isolated in Exercise 4.

E. coli Chromosomal DNA: Isolated in Exercise 2.

Bacteriophage lambda DNA: See Appendix 2.

Water (ddH_2O): Sterile double distilled.

Note: Restriction endonucleases should be diluted to the appropriate concentration before the exercise begins. For most commerical restriction enzymes, a unit of activity is defined as the amount of enzyme required to completely digest 1 μg of pBR322 or λ DNA in 1 hour under the appropriate reaction conditions. The actual rate of DNA cleavage by an enzyme may be influenced by the purity of the DNA sample and the number of cleavage sites relative to those present on the DNA standard. It may be necessary to use more or less than 1 unit/μg DNA to obtain complete digestion in these experiments. The purpose of this exercise is to prepare plasmid and chromosomal DNAs for electrophoretic analysis (Chapter 5) and ligation (Chapter 6). In the process, the digested and undigested restriction patterns of plasmid (miniscreen and purified) and chromosomal DNA will be compared. Molecular weight markers (Day 2) will also be prepared to aid in the electrophoretic characterization of the restriction patterns according to DNA fragment size.

Precautions

1. Before setting up the actual enzyme reactions, practice using the micropipetting apparatus* by dispensing samples of water or buffer into microfuge tubes.

2. A clean pipet tip (or microcapillary) should be used for each transfer. This prevents cross-contamination of the reactants (e.g., DNA with enzyme, enzyme with buffer, and enzyme with ddH$_2$O). The tip should be changed if it touches anything except the reactants and clean glassware.

EXERCISE 6

RESTRICTION ENDONUCLEASE DIGESTION OF PLASMID AND *E. COLI* CHROMOSOMAL DNA

Protocol

DAY 1

1. Using a micropipetter and 1.5 ml microfuge tubes, set up the restriction enzyme digestion reactions indicated below (see fig. 4.3). All reactants and reaction tubes should be held on ice and the reactants added in an ascending order from ddH$_2$O to DNA.

Reaction 4–1	µl
pBR329 miniscreen DNA	2
1 unit *Pst*I endo.	1
10X *Pst*I Buffer	1.5
ddH$_2$O	10.5
	15

Reaction 4–2	µl
pPV33 miniscreen DNA	1
1 unit *Pst*I endo.	1
10X *Pst*I Buffer	1.5
ddH$_2$O	11.5
	15

Reaction 4–3	µl
pPV501 miniscreen DNA	1
1 unit *Pst*I endo.	1
10X *Pst*I Buffer	1.5
ddH$_2$O	11.5
	15

Figure 4.3
Setting up restriction enzyme digestion reactions: (1) Pipetman, (2) microfuge tube, (3) 37°C heating block.

---

*Micropipetting Devices: Two of the most common micropipetters used in research are the Gibson Model P-20D (0 µl–20 µl) and P-200D (0 µl–200µl) Pipetman (West Coast Scientific, Inc.) The popularity of these devices is due to their versatility, accuracy, and reproducibility. A less costly alternative to the Pipetman is the manostat-micropipet filler (Clay-Adams suction apparatus No. 4555), which uses graduated glass microcapillary pipets. Although less versatile than the Pipetman, the manostat-micropipet is accurate, requires no mouth pipetting, and the microcapallary pipets can be sterilized by autoclaving.

EXERCISE 6

RESTRICTION ENDONUCLEASE DIGESTION OF PLASMID AND E. COLI CHROMOSOMAL DNA

2. When the pBR329 miniscreen DNA is added to Reaction 4-1, vortex the contents gently for 2 seconds and immediately dispense 7 µl of this reaction into a microfuge tube containing 3 µl of Reaction Stop Mix. This tube represents the undigested control sample for the miniscreen DNA.

3. Transfer Reactions 4-1, 4-2, and 4-3 from ice to 37°C and incubate for 30 minutes. Terminate the reactions by adding 3µl of Reaction Stop Mix to Reaction 4-1 and 5 µl of the same mix to Reactions 4-2 and 4-3.

4. As described above, set up the following digestion reactions for the purified DNAs of plasmids pBR329, pPV33, and pPV501.

Reaction 4-4

	µl
3 µg pure pBR329 DNA	6
3 units *Pst*I endo.	3
10X *Pst*I Buffer	2
ddH$_2$O	9
	20

Reaction 4-5

	µl
0.5 µg pure pPV33 DNA	1
1 unit *Pst*I endo.	1
10X *Pst*I Buffer	1
ddH$_2$O	7
	10

Reaction 4-6

	µl
0.5 µg pure pPV501 DNA	1
1 unit *Pst*I endo.	1
10X *Pst*I Buffer	1
ddH$_2$O	7
	10

Reaction 4-7

	µl
3 µg pure pBR329 DNA	6
3 units *Hind*III endo.	3
10X *Hind*III Buffer	2
ddH$_2$O	9
	20

Note: The volumes of DNA added to these reactions are based on a presumed DNA concentration of 0.5 µg/µl. Plasmid DNA should be diluted or concentrated accordingly. As a general rule, the volume of DNA added should not exceed 50% of the total reaction volume.

5. After the purified pBR329 DNA has been added to Reactions 4-4 and 4-7, vortex gently for 2 seconds and immediately dispense 1 µl from Reaction 4-4 into a microfuge tube containing 9 µl of Reaction Stop Mix. This tube represents the undigested control sample for Reaction 4-4.

6. Transfer Reactions 4-4 and 4-7 from ice to 37°C and incubate for 1 hour. Place these tubes in the freezer (−20°C) to terminate the restriction enzyme digestion. However, before placing the reactions at −20°C, remove 1 µl from each tube and dispense into 2 microfuge tubes each containing 9 µl of Reaction Stop Mix. These samples will be used to monitor the degree of digestion by gel electrophoresis (Chapter 5).

7. Transfer Reactions 4-5 and 4-6 from ice to 37°C and incubate for 30 minutes. Terminate these reactions by the addition of 10 µl of Reaction Stop Mix to each tube.

8. Store all samples containing Reaction Stop Mix at 4°C, until they can be examined electrophoretically.

DAY 2

1. Set up the following restriction enzyme digestions using the *E. coli* (AB257) chromosomal DNA prepared in Exercise 2 and high molecular weight bacteriophage λ DNA (see Appendix 2).

Reaction 4-8	μl		Reaction 4-9	μl		Reaction 4-10	μl
4 μg *E. coli* DNA	8		4 μg *E. coli* DNA	8		10 μg λ DNA	20
4 units *Hin*dIII endo.	4		4 units *Pst*I endo.	4		5 units *Hin*dIII endo.	5
10X *Hin*dIII Buffer	2		10X *Pst*I Buffer	2		10X *Hin*dIII Buffer	5
ddH$_2$O	6		ddH$_2$O	6		ddH$_2$O	20
	20			20			50

Note: The volumes of DNA added to these reactions are based on a presumed DNA concentration of 0.5 μg/μl. *E. coli* and λ DNA should be concentrated or diluted accordingly. See Table 4.3 for a summary of enzyme reactions.

2. After the *E. coli* DNA has been added to the other components, vortex the contents of each tube gently for 2 seconds and dispense 1 μl from Reactions 4–8 and 4–9 into two separate microfuge tubes containing 9 μl of Reaction Stop Mix. The tubes represent the undigested control samples for the chromosomal DNA.

EXERCISE 6

RESTRICTION ENDONUCLEASE DIGESTION OF PLASMID AND *E. COLI* CHROMOSOMAL DNA

Table 4.3
Summary of restriction enzyme digestion reactions

REACTION	DAY 1							DAY 2		
	4-1	4-2	4-3	4-4	4-5	4-6	4-7	4-8	4-9	4-10
DNA	pBR329[a]	pPV33[a]	pPV501[a]	pBR329[b]	pPV33[b]	pPV501[b]	pBR329[b]	*E. coli*	*E. coli*	λ
Enzyme	*Pst*I	*Pst*I	*Pst*I	*Pst*I	*Pst*I	*Pst*I	*Hin*dIII	*Hin*dIII	*Pst*I	*Hin*dIII
Reaction Volume	15	15	15	20	10	10	20	20	20	50
Incubation Time	30 min.	30 min.	30 min.	1 hr.	30 min.	30 min.	1 hr.	1 hr.	1 hr.	2 hr.
Terminated	3 μl RSM[c]	5 μl RSM	5 μl RSM	−20°C	10 μl RSM	10 μl RSM	−20°C	−20°C	−20°C	−20°C
Final Volume	11 μl	20 μl	20 μl	18 μl	20 μl	20 μl	19 μl	18 μl	18 μl	49 μl
Control Sample	7 μl	—	—	1 μl	—	—	—	1 μl	1 μl	—
Monitoring Sample	—	—	—	1 μl	—	—	1 μl	1 μl	1 μl	1 μl

[a] Miniscreen DNA
[b] Purified DNA (Large-Scale Preparation)
[c] Reaction Stop Mix

EXERCISE 6

RESTRICTION ENDONUCLEASE
DIGESTION OF PLASMID AND
E. COLI CHROMOSOMAL DNA

3. Transfer the three reaction tubes from ice to 37°C and incubate Reactions 4-8 and 4-9 for 1 hour, and Reaction 4-10 for 2 hours. Place these tubes in the freezer to terminate the enzyme reactions. Before freezing, remove 1 µl of each reaction and place it into microfuge tubes containing 9 µl of Reaction Stop Mix. These tubes (stored at 4°C) will be used in Exercise 7 to monitor the progress of each reaction.

CHAPTER 5

Gel Electrophoresis

The construction and isolation of a recombinant DNA molecule provide a convenient method of obtaining specific DNA fragments in amounts sufficient for detailed characterization. A wide variety of techniques have been used to characterize recombinant DNA molecules, including but not limited to velocity sedimentation, measurement of buoyant density, hybridization analysis, electron microscopy, and analytical gel electrophoresis. With the introduction of agarose and polyacrylamide gels to nucleic acid research, electrophoretic techniques have become the principal means for analyzing and characterizing recombinant DNA molecules. This section will deal with the two analytical techniques most commonly used in the preliminary characterization of DNA molecules: gel electrophoresis and hybridization analysis.

Principles of Electrophoresis

When a molecule is placed in an electric field, it will migrate to the appropriate electrode at a velocity, or electrophoretic mobility, proportional to the field strength and the net charge on the molecule. Since a nonreactive, stable medium is used to reduce convectional motion, the electrophoretic mobility is inversely proportional to the frictional coefficient of a molecule (which is a function of the size and shape of the molecule and the viscosity of the medium). Therefore, a mixture of different molecules can be resolved electrophoretically on the basis of: (1) the size of the molecule, which for most biological macromolecules refers to mass or length; (2) the shape or conformation of the molecule; and (3) the magnitude of the net charge on the molecule. When placed at the same position in an electric field, these molecules will resolve into bands and migrate at different rates to different positions in the medium.

Under physiological conditions, the phosphate groups in the phospho-sugar backbone of nucleic acids are ionized. Polynucleotide chains of DNA and RNA are called "polyanions" and they will migrate to the positive electrode (anode) when placed in an electric field. Due to the repetitive nature of the phospho-sugar backbone, double-stranded nucleic acids have roughly the same net charge to mass ratio and will migrate to the anode at equal velocities.

By adjusting the viscosity of the medium, however, the effects of friction and molecular shape can be emphasized to enable nucleic acids to be separated electrophoretically by size. The viscosity of the support medium can be determined by adjusting the degree of crosslinking in the porous matrix of polyacrylamide and agarose gels. Since DNA molecules must move through the pores of the matrix as they migrate toward the anode, their velocity is profoundly influenced by the size of both the molecule and the pore.

Polyacrylamide gels are generated by crosslinking polymers of acrylamide with the co-monomer, *bis*-acrylamide. The covalent crosslinking reaction is generally catalyzed by ammonium persulfate and initiated by TEMED, a tertiary amine. A wide range of pore sizes can be obtained by varying: (1) the concentrations of acrylamide and *bis*-acrylamide, and (2) the degree of polymerization (chain length). Agarose, a linear polysaccharide of galactose and a galactose derivative, is crosslinked by hydrogen bonding. In this instance, the pore size in the gel matrix can be varied by adjusting the concentration of agarose from approximately 0.5% to 1.5%. Since a polyacrylamide matrix has a smaller pore size than an agarose matrix, acrylamide gels are generally used to resolve DNA fragments of less than 4 kilobase pairs, while molecules as large as 200 kilobase pairs can be resolved on agarose gels. As a general rule, a 1% agarose gel of 12 cm pathlength can resolve linear DNA fragments in the range of 30 to 0.2 kilobase pairs, whereas a 7.5% polyacrylamide gel of the same length can resolve fragments in the range of 2 to .05 kilobase pairs in size (fig. 5.1). It has been demonstrated that the

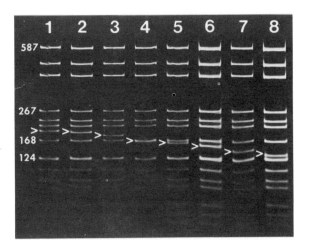

Figure 5.1
Resolving power of a 7.5% polyacrylamide gel. *Hae*III restriction enzyme digestion pattern of pBR327 and seven plasmid derivatives carrying deletions in the promoter for the Tc^r gene. Lane 1—pBR327, arrow indicates the 192 bp *Hae*III fragment carrying the wildtype Tc^r promoter; Lane 2—pPV35, 5 bp deletion; Lane 3—pPV34, 13 bp deletion; Lane 4—pPV32, 29 bp deletion; Lane 5—pPV33, 32 bp deletion; Lane 6—pPV31, 39 bp deletion; Lane 7—pPV37, 49 bp deletion; and Lane 8—pPV36, 60 bp deletion.

electrophoretic mobilities of DNA fragments are a linear function of the log of their molecular weights (for a size range of 20–1,000 base pairs). This relationship is illustrated in Fig. 5.2. It has been recently proposed that this relationship holds for larger DNA molecules when their molecular weights are raised to the ($-2/3$) power.

CHAPTER 5

GEL ELECTROPHORESIS

The dye ethidium bromide fluoresces when irradiated with UV light of a wavelength of 300 nm. Since ethidium bromide can intercalate between the bases of RNA and DNA, it can be used as a means of visualizing nucleic acids in agarose and polyacrylamide gels. When a gel containing molecules of DNA is soaked in a solution of ethidium bromide, the dye will bind to all available sites on the DNA. Since the dye does not bind to agarose or acrylamide polymers, DNA molecules can be visualized as

Visualizing DNA Molecules by Fluorescence

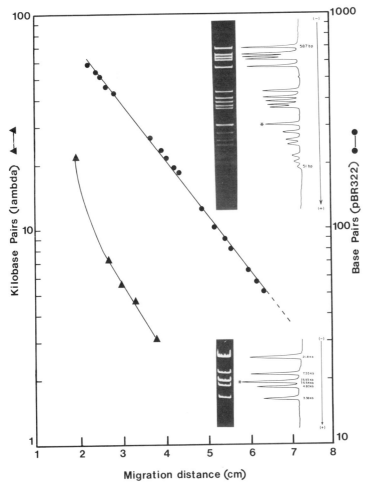

Figure 5.2
Electrophoretic migration pattern of double-stranded DNA fragments of known molecular weight. Triangles show the log-molecular weight/migration plot for the *Eco*RI restriction fragments of bacteriophage lambda DNA on a 1% agarose gel. The circles represent a similar plot of the *Hae*III restriction fragments of pBR322 run on a 7.5% polyacrylamide gel. The asterisk on the densitometer scan indicates comigrating DNA fragments (doublets) commonly observed with lambda and pBR322 DNA.

CHAPTER 5
GEL ELECTROPHORESIS

fluorescent orange bands when the stained gel is irradiated with UV light. Furthermore, under the appropriate staining conditions, the intensity of fluorescence is directly proportional to the size (or mass) of the DNA fragments. In an electrophoretic pattern involving several DNA fragments, the intensity of fluorescence of each band should decrease from the largest to the smallest fragment in a manner that is proportional to fragment size. This is an important feature of the ethidium bromide-gel electrophoresis system in that it enables the investigator to identify comigrating DNA fragments and partial restriction endonuclease digestions by comparison of the relative intensities of the stained bands with molecular weights of the DNA fragments.

Interpreting Electrophoretic Patterns

Because the migration of a molecule is inversely proportional to its frictional coefficient (which is dependent on the size and shape of the molecule), the electrophoretic analysis of DNA can be complicated by the variety of conformations that a DNA molecule can assume. For example, a plasmid DNA molecule such as pBR322 can exist as three different conformational isomers: supercoiled (Form I), relaxed circle (Form II), and linear (Form III) (fig. 5.3). In terms of their relative mobilities through the gel matrix, Form II DNA would be expected to be the slowest since it

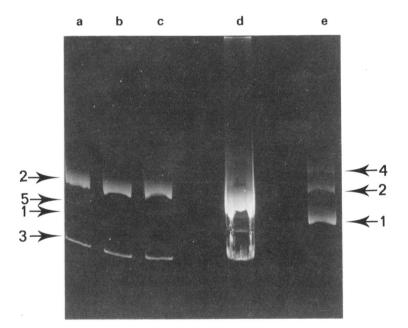

Figure 5.3
Vertical slab gel (1%) electrophoresis of undigested and partially HindIII digested pBR322 DNA. Lanes a, b, and c, designed to show edge effects, also show the different conformational forms of pBR322 (numbered arrows). Arrow 1—supercoil monomer (form I); Arrow 2—relaxed or open circle monomer (form II); Arrow 3—unit length linear (form III); Arrow 4—supercoiled dimer; Arrow 5—dimer linear. Relaxed dimer circles are not visible on this gel. Lane d, which contains 10-fold (2 μg) more plasmid DNA than the other lanes, is designed to show effects of overloading a sample well. Lane e contains the undigested pBR322 control sample.

is not as compact as Form I and cannot "snake" its way through the gel matrix like Form III. Under the electrophoretic conditions used in this manual, the relative mobilities of the three forms of pBR322 (and pBR329) are as follows: Form III (linear) > Form I (supercoiled) > Form II (relaxed circle). It should be stressed, however, that this relationship is only constant for a given set of circumstances. The electrophoretic mobilities of the different forms of DNA are easily influenced by pore size, current, molecular weight, and buffer system. For example, in a 1% agarose gel run in Tris-acetate-EDTA, the relative mobilities of the three forms of pBR322 are: Form I > Form III > Form II. Furthermore, if electrophoresis is performed in the presence of an intercalating dye such as ethidium bromide, the superhelical twist induced in the relaxed circular DNA by the binding of the dye will convert the relaxed circle to a supercoiled molecule. The migration rate of the ethidium-saturated supercoil molecule will now be increased to the migration rate of the unsaturated-supercoiled DNA molecule. The electrophoretic analysis of plasmid DNA can be further complicated by the presence of relaxed and supercoiled "dimers" (circular concatamers of two plasmid molecules) in the preparation of plasmid DNA. Generally speaking, these forms will migrate slightly slower than their respective monomeric forms.

Gel Systems

While the electrophoresis of DNA can be performed in tubular gels, the slab gel is preferable since it can accommodate many different samples and the electrophoretic conditions for each sample are relatively constant across the gel. The principles of vertical and horizontal slab gel electrophoresis are similar, with each system having applications for which it is better suited than the other. Vertical gel electrophoresis, which can be performed using agarose or acrylamide, is technically easy, gives reproducible results, and can be completed more rapidly than many horizontal systems (fig. 5.4a–c). Horizontal agarose gel electrophoresis has the advantage that much lower concentrations of agarose can be utilized than in a vertical gel system, allowing the separation of large DNA fragments.

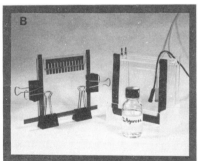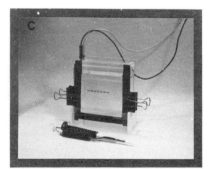

Figure 5.4a–c
Vertical slab gel electrophoresis apparatus and its assembly. (a) Components; 1—two 6½" × 6½" glass plates, one notched; 2—lower buffer chamber; 3—upper buffer chamber; 4—spacers and comb; 5—large paper clips. (b) Casting the slab gel and assembling the upper and lower chambers. Before casting agarose and polyacrylamide gels, make sure the assembled plates and spacers are sealed with 2 ml of molten agarose. (c) Apparatus is loaded with DNA samples and ready to run.

In addition, horizontal electrophoresis can be performed in the presence of ethidium bromide, allowing the visualization of the DNA fragments during electrophoresis with the aid of a hand-held ultraviolet light. Both types of slab gel electrophoresis have distinct advantages over the more conventional tube gel electrophoresis.

An effective compromise between the standard vertical and horizontal slab gels is the "submarine minigel" (fig. 5.5a–c). These small (6 cm × 4 cm × 0.2 cm) agarose slab gels are run horizontally under approximately 1 mm buffer and have the advantages of being easier to prepare and faster-running than conventional gels. The *Eco*RI restriction fragments of bacteriophage λ, for example, can be resolved on a minigel in 20 minutes as compared to 1.5 hours and 12 hours for standard vertical and horizontal gels, respectively. This faster rate of migration is due to the greater conductance (i.e., greater ability to conduct current) of the submarine gel system. If the same level of current were applied to a vertical or horizontal gel, the resistance (the reciprocal of conductance) in these gels would overheat the system, causing distortion of the electrophoretic pattern. While the convenience and speed of the minigel system make it a useful tool for routine characterizations of DNA, its resolving power is less than that of the more conventional gel systems. Therefore, minigels are not recommended for precise electrophoretic analysis of DNA molecules.

Gel Artifacts

Gel electrophoresis is susceptible to a number of artifacts that can affect the interpretation of results. These artifacts are often due to a failure to maintain a constant electric field throughout the slab gel. The most common cause of field strength variation is due to unequal dissipation of heat (a function of resistance) from the gel. As the voltage applied to a gel is increased, the amount of heat generated during electrophoresis increases proportionately. As the rate of heat production exceeds the ability of the system to dissipate the heat, the result is a net increase in the temperature of the gel. Because the heat dissipating capacity of the slab gel is greater at the edges adjacent to the nonconducting spacer material, the edges will

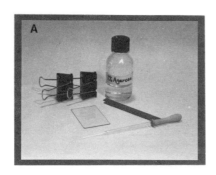 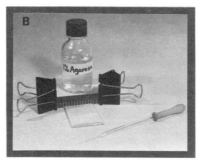 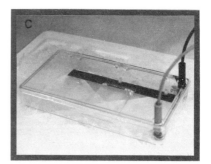

Figure 5.5a–c
Submarine minigel electrophoresis apparatus and its assembly. (a) Components; 1—one 6 cm × 4 cm microscope slide; 2—well-forming comb. (b) Casting the minigel. (c) Apparatus is loaded with DNA samples and ready to run. Note that the apparatus is sitting in a tray of ice.

remain cooler than the interior regions of the gel. As the conductivity of the gel increases with temperature increase, the electrophoretic current at the edges will be slightly less than the current in the warmer center of the gel. When identical DNA samples are applied to the edge and the center of a slab gel and the gel operated at a power that exceeds its heat dissipating capacity, the sample in the center will migrate more rapidly than the sample at the edge (fig. 5.3, lane 1). Edge effects can complicate the comparison of samples in various regions of a gel. The use of an electrophoretic apparatus that optimizes heat dissipation and the application of appropriate voltage levels to gels can greatly reduce the occurrence of edge effects.

Another common problem associated with slab gel electrophoresis involves the effects of the ionic strength on the migration rate of the sample. Increasing the ionic strength of a sample tends to decrease the migration rate of low molecular weight molecules during electrophoresis. For example, plasmid DNA can be digested with a restriction enzyme to generate a characteristic set of DNA fragments. If four aliquots of these fragments are taken and adjusted with sodium chloride to 0.05, 0.10, 0.15, and 0.20 M NaCal, the effect of ionic strength on electrophoretic mobility can be observed. As shown in Fig. 5.6, when these aliquots are subjected to agarose slab gel electrophoresis along with the original DNA fragments, the increasing concentration of salt retards the migration of the smaller DNA fragments.

An estimate of the size of a recombinant plasmid is generally obtained by digesting the plasmid DNA with a restriction endonuclease prior to gel electrophoresis. For example, if DNA is inserted into the *Sal*I

CHAPTER 5
GEL ELECTROPHORESIS

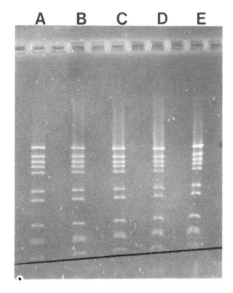

Figure 5.6
An agarose gel (1%) showing the effects of increasing salt on the mobility of the *Hin*dIII restriction fragments of *Bacillus* phage, φ29. Lane A—no salt; Lane B—50 mM NaCl; Lane C—100 mM NaCl; Lane D—150 mM NaCl; Lane E—200 mM NaCl. The black line illustrates the retarding effect of salt on the smaller restriction fragments.

cleavage site of the plasmid pBR322, digestion of the resulting recombinants with *Sal*I will release the inserted DNA fragments. Separation of the fragments by slab gel electrophoresis will identify the pBR322 cloning vector and the cloned DNA fragments. When molecular weight standards are included, molecular weight estimates for the cloned DNA fragments can be obtained by constructing the semilog plot of the two samples, as shown in Fig. 5.1. When the salt concentration present in the standards is significantly different from that in the samples, these molecular weight estimates can be in error by as much as 15%. As the reaction conditions for different restriction enzymes may involve different salt concentrations, the error resulting from ionic variation of samples can be minimized by adjusting the salt concentrations of the samples to a common molarity after enzymatic digestion but prior to electrophoresis.

Tracking Dye

A good tracking dye solution or "stop mix" is vital to the success of an electrophoretic experiment. The stop mix serves three important functions: (1) it is used to terminate enzymatic reactions, (2) it provides a convenient means to load samples on the gel, and (3) it provides a way to monitor the progress of the experiment so that electrophoresis can be stopped at the appropriate time. For this reason, most tracking dye solutions contain an agent such as SDS or urea to denature proteins, ficoll or glycerol to provide viscosity to the sample, and one or more nonintercalating dyes with specific electrophoretic mobilities. Bromphenol-blue (dark blue) and xylene cyanole FF (turquoise) are two dyes commonly used with agarose and polyacrylamide gel electrophoresis. Under the conditions described in this manual, these dyes migrate approximately with 30 base pair and 4,000 base pair DNA fragments, respectively. Therefore, these dyes can be used to estimate the rate of migration and positon of a DNA fragment in the gel.

Transfer-Blot/Hybridization Analysis

By using gel electrophoresis in combination with restriction endonuclease analysis, a detailed and accurate physical map of a DNA molecule or chromosome can be constructed. However, in order to construct a genetic map indicating the size and location of genetic coding regions on this molecule, more information is required. This information can be obtained by combining agarose gel electrophoresis with hybridization analysis by means of the blotting method originally described by E. Southern.

The fact that complementary single strands of DNA and RNA can anneal or hybridize to generate a homoduplex of DNA-DNA or heteroduplex of DNA-RNA has been well established. The observation that this hybridization can take place even when one of the two complementary strands is immobilized on a nitrocellulose filter provides the conceptual basis for the "Southern blot" technique (fig. 5.7). As illustrated in Fig. 5.7, restriction DNA fragments are denatured in the gel and blotted onto a sheet of nitrocellulose by elution or electrophoretic transfer in a manner that does not disturb their original pattern. After the single-stranded DNA is permanently bound to the nitrocellulose, the sheet is incubated in a solution containing radioactively labeled "probe" (i.e., complementary DNA or RNA). Once the homologous sequences have had time to anneal, the nitrocellulose is washed free of unhybridized probe and placed in contact with X-ray film. The resulting autoradiograph will indicate which restriction fragments bear homology to the nucleotide sequence on the probe. The autoradiographic pattern can be compared to the ethidium bromide staining pattern to determine which restriction fragments correspond to coding regions in the DNA. If the hybridization probe consists of a cloned DNA fragment corresponding to the wildtype version of a particular gene, the blot-transfer hybridization analysis can be used to determine the nature and extent of mutations in that gene. For example, if a mutation is of the insertion or deletion type, the autoradiographic pattern produced from a hybridization between wildtype probe and wildtype DNA should be distinctly different from that produced

CHAPTER 5

GEL ELECTROPHORESIS

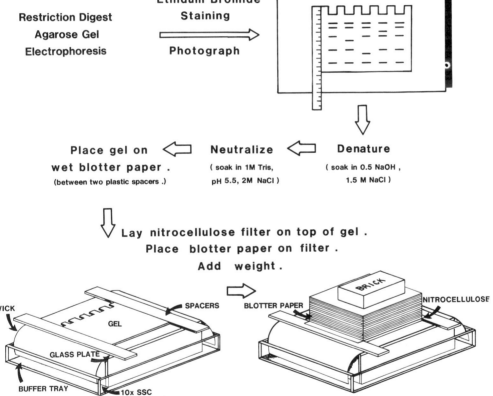

Figure 5.7

when wildtype probe and mutant DNA are used. These examples serve to illustrate how the usefulness of agarose gel electrophoresis can be extended by the Southern blot technique to enable physical and genetic maps of DNA molecules to be constructed almost simultaneously.

Finally, the concept of blotting itself has been extended to include RNA and protein. These blotting procedures, now known in the vernacular as "Northern" and "Western" blotting, respectively, differ from the Southern techniques in two principal ways. Northern blots involve the blotting of RNA onto diazotized paper where the RNA becomes permanently bound by covalent crosslinking. With Western blotting, proteins are separated electrophoretically, blotted and bound to nitrocellulose filters, and then challenged with radioactively labeled (^{125}I) antibody to one particular protein. Both procedures are useful in establishing the presence and approximate size of a particular species of RNA or protein within the cell.

References

Bearden, J. C. 1979. Electrophoretic mobility of high-molecular-weight, double-stranded DNA on agarose gels. Gene 6:221–234.

Bittner, M., Kupferer, P., and Morris, C. F. 1980. Electrophoretic transfer of protein and nucleic acid from slab gels onto diazobenzyloxymethyl paper or nitrocellulose sheets. Anal. Biochem. 102:459–471.

Bøvre, K., and Szybalski, W. 1971. Multistep DNA-RNA hybridization techniques. In Methods in Enzymology, L. Grossman and K. Moldave, eds., 21:350–383.

Bowen, B., Steinberg, J., Laaemli, U. K., and Weintraub, H. 1980. The detection of DNA-binding proteins by protein blotting. Nucl. Acids Res. 8(1):1–20.

Cooper, T. G. 1977. The Tools of Biochemistry. Wiley. New York.

Maniatis, T., Jeffrey, A., and van deSande, H. 1975. Chain length determination of small double- and single-stranded DNA molecules by polyacrylamide gel electrophoresis. Biochem. 14:3787–3794.

Renart, J., Reiser, J., and Stark, G. R. 1979. Transfer of proteins from gels to diazobenzyloxymethyl paper and detection with antisera. Proc. Natl. Acad. Sci. USA 76(7):3116–3120.

Sharp, P. A., Sugden, B., and Sambrook, J. 1973. Detection of two restriction endonuclease activities in *Haemophilus parainfluenzae* using analytical agarose-ethidium bromide electrophoresis. Biochem. 12:3055–3063.

Southern, E. 1975. Detection of specific sequences among DNA fragments separated by gel electrophoresis. J. Mol. Biol. 98:503–517.

Southern, E. 1979. Gel electrophoresis of restriction fragments. In Methods in Enzymology, R. Wu, ed., 68:152–176.

Studier, F. W. 1973. Analysis of bacteriophage T7 early RNAs and proteins on slab gels. J. Mol. Biol. 79:237–248.

Towbin, H., Staehelin, T., and Gordon, J. 1979. Electrophoretic transfer of proteins from polyacrylamide gels to nitrocellulose sheets; procedure and some applications. Proc. Natl. Acad. Sci. USA 76(9):4350–4354.

EXERCISE 7

Agarose Gel Electrophoresis of Plasmid and Chromosomal DNA

Materials

Tris-Borate Buffer (1X): 90 mM Tris-HCl pH 8.2, 2.5 mM Na_2 EDTA, 89 mM boric acid. (1X Tris-Borate Buffer is made from a 10X stock solution; see Appendix 1.)

Reaction Stop Mix: 5 M urea, 10% glycerol, 0.5% SDS, 0.025% xylene cyanole FF, 0.025% bromphenol-blue WS.

Agarose: Molten, 1% solution made in Tris-Borate Buffer (1X).

Ethidium Bromide: 4 µg/ml, made with water.

Phenol: Equilibrated with an equal volume of 50 mM Tris-HCl pH 7.5, 100 mM NaCl, 1 mM NaN_3.

Ether: Hydrated with one-tenth volume water.

Chloroform.

TEN Buffer.

Protocol

DAY 1

1. Set up a submarine minigel or vertical slab 1% agarose gel as shown in Figs. 5.4 and 5.5. The gel should have at least 12 sample-loading wells.

2. Once the gel is submerged in 1X Tris-Borate Buffer, load samples just above the surface of the sample-loading wells. Load the reactions and volumes indicated below:

					—Gel 1—							
Well No.:	1	2	3	4	5	6	7	8	9	10	11	12
Reaction:			4–1		4–2	4–3		4–5	4–6			
Control Sample:				4–1								
Monitoring Sample:							4–10					
µl loaded: (Vert. slab)			10	10	20	20	10	20	20			
µl loaded: (minigel)			10	10	10	10	5	10	10			

EXERCISE 7

AGAROSE GEL ELECTROPHORESIS OF PLASMID AND CHROMOSOMAL DNA

Note: The monitoring sample of Reaction 4–10 is being run on Gel 1 to determine the extent of the *Hin*dIII digestion. If the reaction is complete, the *Hin*dIII cleaved λ DNA will serve as molecular weight standards from which the size of neighboring DNA molecules can be determined. Before loading, heat the λ DNA to 65°C for 5 minutes to dissociate the cohesive-overlapping strands (cos) on the 23.7 kb and 4.26 kb *Hin*dIII fragments. (See the inside cover for a chart of common molecular weight standards.) The samples have been loaded in the center of the gel to avoid edge effects, and *Hin*dIII digested DNA was loaded asymmetrically to the right. Since gels can be inadvertently flipped during staining and photographing, the asymmetry of the molecular weight standards enables the left and right sides of the gel to be distinguished.

3. After the samples have been loaded, apply power to the electrophoresis chamber (be sure that both electrode wires are submerged). Run the vertical slab gel at 150 volts (25–30 milliamps) for 80 minutes. Minigels should be run at 100 volts (10–15 milliamps) for 30 minutes. Monitor the progress of the fast-running (bromphenol blue) tracking dye during electrophoresis. For most experiments, this dye should not run out of the gel.

4. When electrophoresis is complete, turn off the power supply, disconnect the positive and negative terminals, and transfer the gel to a staining tray containing 200 ml of ethidium bromide solution. (*Caution*: Ethidium bromide is a potent mutagen. Wear gloves at all times when handling gels in this solution. Decontaminate surface spills with a 20% solution of bleach.)

5. After gels have been stained from 1 to 5 minutes in the ethidium bromide solution, transfer the gel (with the aid of a kitchen-type spatula) to a tray of water where the gel can destain for about 5 minutes.

6. Using a Polaroid MP3 or MP4 camera and a UV transilluminator (Ultra-Violet Products Inc.—Chromato-Vue Transilluminator Model C-61), the stained gel can be photographed through a Wratten No. 9 filter (to reduce near-UV light) onto Polaroid Type 55 film. The negative produced by this film should be hardened by a dipping treatment in an 18% solution of sodium sulfite. The negative should be washed in water for 5 minutes and dried. (*Caution*: Use UV resistant eye protection when photographing gels.)

7. If the λ DNA has been digested to completion (8 fragments), the main pool of reaction 4–10 should be thawed and immediately terminated by phenol/chloroform extraction.

8. To the remaining 49 µl of Reaction 4–10, add 100 µl of buffer saturated phenol and 100 µl of chloroform. Vortex the mixture vigorously for 1 minute. Separate the phases by microcentrifugation. Transfer the upper (aqueous) phase to a clean microfuge tube.

9. Add 300 µl of hydrated ether and vortex the mixture vigorously for 1 minute. Remove the upper ether phase and allow residual ether to evaporate. Adjust the concentration of DNA to approximately 0.25 µg/ml by the addition of TEN buffer. Store at 4°C until further use.

10. If Reaction 4-10 did not go to completion (i.e., more than 5 restriction fragments are visible on the gel), add one-half the original volume of enzyme and incubate the mixture another hour.

EXERCISE 7

AGAROSE GEL ELECTROPHORESIS OF PLASMID AND CHROMOSOMAL DNA

DAY 2

1. Load onto a 1% agarose gel the reactions and volumes indicated below:

—Gel 2—

Well No.:	1	2	3	4	5	6	7	8	9	10	11	12
Reaction:		—	—	—	—	—	—	—	—	—		
Control Sample:		4-4		4-7		4-8			4-9			
Monitoring Sample:			4-4		4-7		4-8	4-10		4-9		
µl loaded: (Vert. slab)		10	10	10	10	10	10	10*	10	10		
µl loaded: (minigel)		10	10	10	10	10	10	5	10	10		

*1 µl HindIII digested λDNA (0.25 µg/µl) in 9 µl Reaction Stop Mix.

2. Electrophorese samples, and stain and photograph gel as described on Day 1, steps 3 through 6.

3. If all samples have been digested to completion, terminate the reactions by heat inactivation (65°C, 5 minutes) and store frozen until they are ready to be ligated (Chapter 6).

4. If the reactions did not go to completion, add one-half the original volume of endonuclease and incubate for another hour.

CHAPTER 6

Ligation of DNA

The construction of recombinant DNA molecules is dependent on the ability to covalently seal single-strand breaks or nicks in DNA. This process is accomplished both *in vivo* and *in vitro* by the enzyme polynucleotide (DNA) ligase. This enzyme catalyzes the formation of a phosphodiester bond between the 3' hydroxyl and the 5' phosphate of two adjacent nucleotides, thus reestablishing the structural continuity of the DNA strand (fig. 6.1a). Because of this ability to repair single- or double-strand scissions in DNA, DNA ligase plays a critical role in the processes of recombination, DNA replication, and the repair of DNA damage. The ability to join blunt- or staggered-end DNA fragments also makes DNA ligase a valuable tool in the arsenal of recombinant DNA techniques.

DNA Ligase from *E. coli* and Bacteriophage T4

DNA ligases have been isolated from a variety of prokaryotic and eukaryotic organisms and their viruses. All were found to be capable of catalyzing the reaction depicted in Fig. 6.1a. With *E. coli* DNA ligase (a 74,000 dalton protein), NAD^+ is apparently involved in the formation of an enzyme-adenylate complex prior to the formation of the phosphodiester bond. With bacteriophage T4 DNA ligase, ATP functions in the formation of the enzyme-adenylate complex. In both cases, the adenylyl residue is transferred to the 5' phosphate of the nick being repaired to generate Ad-P-P-DNA. The formation of the phosphodiester linkage to the 3' hydroxyl DNA strand proceeds with the release of NMN or AMP.

The bacteriophage T4 codes for its own DNA ligase. This enzyme, the product of gene 30, has a molecular weight of approximately 60,000 daltons and requires ATP as an energy source. Unlike DNA ligase purified from *E. coli*, T4 DNA ligase has the unique ability to join together flush or blunt-end DNA fragments in a process referred to as blunt-end

CHAPTER 6
LIGATION OF DNA

ligation (fig. 6.1b). Although the original source of this enzyme was T4-infected cells of *E. coli*, the purification of T4 DNA ligase has been greatly simplified by the construction of *E. coli* strains lysogenic for a recombinant phage lambda containing the T4 DNA ligase gene. Growth of these strains at a restrictive temperature (42°C) results in the overproduction of T4 DNA ligase. Since high levels of ligase are present in the thermally induced cells, purification procedures have been developed that enable large quantities of high specific activity enzyme to be isolated easily and rapidly. Because of the availability of T4 DNA ligase and its blunt-end ligation capability, this is the enzyme of choice for the *in vitro* ligation of DNA fragments.

In Vitro and *In Vivo* Ligation of DNA Fragments

In its simplest form, the construction of a recombinant plasmid involves a bimolecular reaction in which one end of a linearized plasmid vector is ligated to a target DNA fragment, after which the linear chimera is circularized by the ligation of the two remaining ends. The circularization process can be accomplished by either of two means: the ligation can be performed inside the bacterial cell, or *in vitro*, prior to the transformation of the bacteria with the recombinant molecules. When using the ligation capabilities of the bacterial cell, the DNA molecules must possess complementary overlapping termini of appropriate length and base composition so that the ends of the DNA molecules can anneal under physiological conditions. This can be achieved by cleaving the DNA fragments with restriction enzymes such as *Eco*RII (↓CC(^A_T)GG) or by using N-terminal transferase to add homopolynucleotide extensions to the DNA fragments (see Chapter 1). Once in the cell, the complementary termini anneal and the cell ligates the nicks, thus generating the desired recombinant molecule. Although this method has been used in the past, the over-

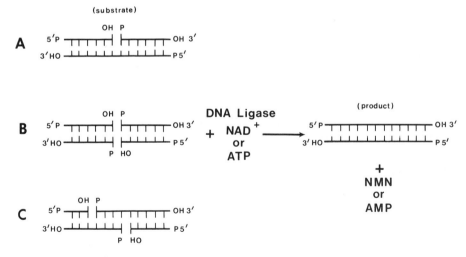

Figure 6.1
Action of DNA ligase on (a) nicked, (b) blunt-end, and (c) cohesive-end DNA.

all efficiency of recombinant molecule formation is low and may be accompanied by deletions of nucleotides at the termini.

The use of DNA ligase in the *in vitro* joining of DNA molecules prior to the transformation aids in the construction of recombinant molecules in several ways. *In vitro* ligation of DNA reduces the susceptibility of the DNA to nucleolytic degradation inside the cell and greatly increases transformation efficiencies. Ligation of the cohesive ends generated by restriction endonucleases (fig. 6.1c) will preserve the original cleavage sites, thus facilitating the recovery of cloned DNA fragments. Finally, the ligation reaction conditions can be adjusted in such a manner as to favor the formation of either circular molecules or linear concatemers consisting of several DNA fragments joined end-on-end.

Factors Affecting the Ligation Reaction

The ligation reaction is a process dependent on several parameters, including temperature, ionic concentration, the nature of the DNA ends (cohesive or blunt), the relative concentration of the DNA ends, and the concentration and molecular weight of the DNA fragments. When dealing with DNA fragments with cohesive termini such as those produced by the restriction enzymes *Eco*RI, *Hin*dIII, and *Pst*I, the cohesive ends involve only a few nucleotides. The temperature that disrupts the hydrogen bonding between these ends is very low. For the tetranucleotide AATT produced by the *Eco*RI endonuclease, the temperature at which only one-half of the ends are annealed (T_m) is 5°C. Although T_m values will vary according to the length and base composition of the cohesive ends, the T_m values for most of the cohesive termini generated by restriction endonucleases occur below 15°C. In order to optimize ligation, the reaction should take place under conditions that allow the annealing of the termini. However, the temperature optimum for DNA ligase activity is 37°C, and activity is greatly reduced below 5°C. Therefore, the optimum temperature for the ligation of cohesive termini is a compromise between the T_m of the cohesive ends and the temperature optimum for DNA ligase. A reaction temperature frequently used in the ligation of cohesive ends is 12.5°C.

Two other factors that significantly affect the rate and nature of the ligation reaction are *j*, the effective concentration of two ends of a random coil segment of DNA, and *i*, the total concentration of ends in the ligation reaction. Under a given set of conditions, the value *j* is a constant for each particular type of DNA molecule. While independent of DNA concentration, *j* is a function of DNA length, i.e., molecular weight. The proximity of two ends of a random coil of DNA is influenced by the conformational state of the DNA molecule, which in turn is dependent on ionic strength. An excess of neutralizing cations such as Na^+ make the DNA helix more stable and rigid and decrease the effective concentration of the ends of the molecule (*j*) relative to each other. Since ligation reactions are generally performed in a standard reaction buffer to optimize ligase activity, the ionic influences on the *j* value can generally be ignored in the consideration of ligation reactions.

CHAPTER 6
LIGATION OF DNA

Based on studies of linear bacteriophage lambda DNA, j values for any type of DNA molecule can be calculated relative to the j value of lambda (j_λ):

$$j_x = j_\lambda \left(\frac{mw_\lambda}{mw_x}\right)^{3/2},$$

where

$j_\lambda = 3.6 \times 10^{11}$ ends/ml
$mw_\lambda = 30.8 \times 10^6$ daltons
mw_x is the molecular weight of the unknown DNA molecule
the units of j_x are in ends/ml.

If the cohesive termini of a DNA molecule are self-complementary, then their total concentration i, in ends/ml, is given by:

$$i = 2 N_o M \times 10^{-3}$$

where

N_o = Avogadro's number, 6.022×10^{23} molecules/mole
M = the molar concentration of the DNA molecules.

When j, which describes the effective concentration of one end of a molecule relative to the other end of the molecule, is equal to i, the total concentration of ends, the probability of two ends of the same molecule finding each other should be the same as the probability of finding the end of a different molecule. When j is greater than i, the formation of circular molecules will be favored during ligation. And when j is less than i, the formation of linear concatemers will be favored during ligation.

The j/i ratio for a DNA molecule of a given molecular weight can be calculated according to the formula:

$$j/i = \frac{j_\lambda (mw_\lambda/mw)^{3/2}}{2 N_o \times 10^{-3}}$$

Using the appropriate values for j_λ, mw_λ, N_o, and converting M, the molar concentration of DNA molecules, to [DNA], the DNA concentration in gm/l, this equation can be simplified to:

$$j/i = \frac{51.1}{[DNA] (mw)^{1/2}}$$

A simple rearrangement generates the equation:

$$mw = \left(\frac{51.1}{j/i[DNA]}\right)^2$$

A graph of this expression, shown in Fig. 6.2, can be used to predict the type of products expected during the ligation of a specific DNA fragment at various concentrations. As the ratio j/i decreases, the formation of linear concatamers is favored, and as the ratio j/i increases, the circularization of molecules is favored. For example, when a DNA fragment of 4×10^6 daltons is ligated at a concentration of 10 µg/ml, the ratio $j/i = 5$ indicates that the conditions favor the formation of circular monomers. When the same DNA fragment is ligated at a concentration of 50 µg/ml, the ratio $j/i = 0.5$ indicates that the formation of linear concatamers will be favored. Although theoretical considerations predict that the point of conversion from circularization to concatamerization should occur at $j/i = 1$, experimental observations made by Dugaiczyk et al. indicate that this conversion actually occurs at $j/i = 2-3$.

It should be stressed that a ligation reaction is not an instantaneous event, but rather a progressive series of individual ligation events, where each event significantly alters the j/i ratio for the remaining unligated molecules. For example, if a 1×10^6 dalton DNA fragment is ligated at a concentration of 50 μg/ml ($j/i = 1$), the initial ligation events favor the formation of concatamers. However, when 50% of the ends in the reaction have been ligated into concatameric forms, i has been decreased by a factor of 2, and the ratio j/i for the remaining unligated molecules has thus been increased to 2. When 80% of the ends have been removed by ligation, the j/i for the remaining molecules has been increased to 5. As the reaction proceeds, the ratio j/i increases and favors the formation of circular molecules rather than the linear concatamers.

Although the j/i ratios in Fig. 6.2 apply to the ligation of DNA molecules of the same molecular weight, this graph can be used to estimate reaction conditions for the ligation of DNA fragments of different molecular weights. For example, if the intent of the ligation is to ligate restriction enzyme-digested chromosomal DNA fragments with an average molecular weight of 5×10^6 daltons to an equal amount of linear pBR322 DNA (mw = 2.7×10^6 daltons), a sufficiently accurate approximation can be made by averaging the molecular weights (3.8×10^6 daltons). Thus, a DNA concentration and j/i ratio ($j/i = 0.5$) can be chosen to promote the formation of linear concatamers. This can be achieved by ligating the mixture of DNA fragments at a concentration of 50 μg/ml.

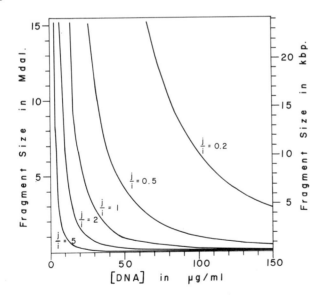

Figure 6.2
A graph of the ratio j/i as a function of DNA fragment size and ligation concentration can be used to predict the products of ligation. For any fragment, as the ligation concentration increases, the corresponding j/i ratio decreases. j/i greater than 1 to 2 favors the formation of circular molecules, while j/i less than 1 to 2 favors the formation of concatameric molecules.

CHAPTER 6
LIGATION OF DNA

When ligating a mixture of DNA fragments, it is also possible to affect the ligation products by manipulating the ratio of the fragments in the ligation reaction. Consider the ligation of a 2×10^6 dalton DNA fragment with a 1×10^6 dalton (target) DNA fragment to form a recombinant molecule. If the ligation is performed at a DNA concentration of 40 µg/ml, $j/i = 1$, and the conditions will favor the formation of concatameric molecules, including molecules in which the two different size fragments have been ligated to each other. However, if the 2×10^6 dalton fragment is present in a 10-fold molar excess over the 1×10^6 dalton fragment, the formation of a dimer of the small fragment will occur at 1/10 of the frequency of the formation of a hybrid molecule consisting of two different size fragments. Therefore, the presence of an excess of the large fragment in the ligation reaction will optimize the formation of multimers of the large fragment and the hybrid molecules while minimizing formation of multimers of the small fragment.

In the construction of a chromosomal DNA gene bank, it is desirable to optimize the formation of recombinant molecules as opposed to nonrecombinant DNA molecules. This can be accomplished by either ligating a 10–20 fold excess of chromosomal DNA to the vector DNA under conditions that favor the formation of concatamers, or by pretreating the plasmid cloning vector with alkaline phosphatase. Both bacterial and calf intestine alkaline phosphatase convert the 5' terminal phosphate residue on a DNA fragment to a 5' terminal hydroxyl. Since DNA ligase requires the presence of a 5' terminal phosphate, cohesive termini can anneal but cannot be ligated in the absence of this moiety. This property can be utilized to prevent recircularization of the plasmid vector on itself by removing the 5' terminal phosphates from the vector DNA after cleavage with a particular restriction enzyme. The plasmid vector can no longer ligate to itself and form either circular monomers or linear concatamers. However, when a second DNA fragment, digested with the same restriction endonuclease, is added to the phosphatase-treated vector DNA, the cohesive termini will anneal, and the 5' terminal phosphate residues provided by the target fragment can serve as substrates for DNA ligase. These cohesive termini will be ligated only in the strand that has the 5' phosphate, leaving the other strand nicked. This ligated DNA can be used to transform bacterial cells, after which nicks are repaired *in vivo*, thus generating viable recombinant plasmids. When using alkaline phosphatase to optimize the formation of recombinant molecules, the best results are obtained with a j/i ratio of 1 and a two-fold excess of vector DNA to target DNA.

Blunt-End Ligation

While the reaction catalyzed by T4 DNA ligase is the same for both cohesive and blunt-end ligations, the lack of cohesive termini makes the latter reaction more complex and significantly slower. Cohesive-end ligations, which are essentially nick-sealing reactions (fig. 6.1c), proceed approximately 100 times faster than the blunt-end ligation. One reason for this is that blunt-end fragments cannot anneal; therefore, the time interval for the juxtaposition of the 5' phosphate to the 3' hydroxyl is exceedingly small. Furthermore, it has been suggested that the blunt-end ligation reaction requires at least two ligase molecules—one to hold the blunt ends in juxtaposition, and one to catalyze the formation of the phosphodiester bond. Therefore, 10 to 30 times more T4 DNA ligase is required to achieve blunt-end ligation rates equivalent to those observed for cohesive-end ligation. Also, since the number of juxtapositions can be increased by increasing the concentration of termini, most blunt-end ligations are described in terms of pmoles or μmoles of ends and frequently use j/i values less than 0.5. Since the thermal stability of annealed cohesive ends is not a consideration with blunt-end ligations, the reaction is less influenced by temperature than the cohesive-end ligation. As a general rule, the optimal temperature for blunt-end ligation should approximate the T_m for the smallest fragment in the reaction, but not exceeding 37°C. As shown in Fig. 6.3, the *Hae*III restriction enzyme cleavage products of pBR322 were successfully ligated at 23°C.

The last feature that distinguishes the two types of ligation reactions is the reversible inhibition of blunt-end ligations by high concentrations of ATP. While a final ATP concentration of 0.5 mM is optimal for both cohesive and blunt-end ligations, ATP concentrations of 2.5 mM and higher selectively inhibit the blunt-end ligating capability of T4 DNA ligase. Therefore, it is important to use the lower ATP concentration for most ligation reactions, unless specific inhibition of blunt-end ligation is desired.

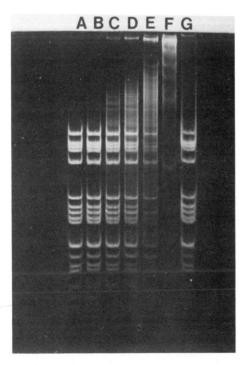

Figure 6.3
The polyacrylamide gel (7.5%) electrophoresis pattern of *Hae*III cleaved pBR322 DNA, incubated with increasing amounts of T4 DNA ligase. One microgram of *Hae*III cleaved pBR322 DNA was incubated at 23°C for 1 hour with the following dilutions of T4 DNA ligase: (A) No ligase. (B) 10^{-3}. (C) 10^{-2}. (D) 10^{-1}. (F) 10^{-0}. (G) Products of Lane F cleaved with *Hae*III endonuclease.

CHAPTER 6
LIGATION OF DNA

References

Dugaiczyk, A., Boyer, H. W., and Goodman, H. M. 1975. Ligation of *Eco*RI endonuclease-generated DNA fragments into linear and circular structures. J. Mol. Biol. 96:171–184.

Ferretti, L., and Sgaramella, V. 1981. Specific and reversible inhibition of the blunt-end joining activity of the T4 DNA ligase. Nucl. Acids Res. 9:3695–3705.

Gellert, M. 1967. Formation of covalent circles of lambda DNA by *E. coli* extracts. Proc. Natl. Acad. Sci. USA 57:148–155.

Heyneker, H. L., Shine, J., Goodman, H. M., Boyer, H. W., Rosenberg, J., Dickerson, R. E., Narang, S. A., Itakura, K., Lin, S., and Riggs, A. D. 1976. Synthetic *lac* operator DNA is functional *in vivo*. Nature 263:748–752.

Higgins, N. P., and Cozzarelli, N. R. 1979. DNA-Joining Enzymes: A Review. In Methods in Enzymology; Recombinant DNA, R. Wu, ed. Volume 68, pp. 50–71. Academic Press Inc., New York.

Murray, N. E., Bruce, S. A., and Murray, K. 1979. Molecular cloning of the DNA ligase gene from bacteriophage T4. II. Amplification and preparation of the gene product. J. Mol. Biol. 132:471–491.

Olivera, B. M., and Lehman, I. R. 1967. Linkage of polynucleotides through phosphodiester bonds by an enzyme from *Escherichia coli*. Proc. Natl. Acad. Sci. USA 57:1426–1433.

Panasenko, S. M., Alazard, R. J., and Lehman, I. R. 1978. A simple, three-step procedure for the large scale purification of DNA ligase from a hybrid λ lysogen constructed *in vitro*. J. Biol. Chem. 253:4590–4592.

Sgaramella, V., and Ehrlich, S. D. 1978. Use of the T4 polynucleotide ligase in the joining of flush-ended DNA segments generated by restriction endonucleases. Eur. J. Biochem. 86:531-537.

Sgaramella, V., and Khorana, H. G. 1972. Studies on polynucleotides: A further study of the T4 ligase-catalyzed joining of DNA at base-paired ends. J. Mol. Biol. 72:493–502.

Sugino, A., Goodman, H. M., Heyneker, H. L., Shine, J., Boyer, H., and Cozzarelli, N. R. 1977. Interaction of bacteriophage T4 RNA and DNA ligase in joining of duplex DNA at base-paired ends. J. Biol. Chem. 252:3987-3994.

Tait, R. C., Rodriguez, R. L., and West, R. W. 1980. The rapid purification of T4 DNA ligase from a λT4*lig* lysogen. J. Biol. Chem. 255:813–815.

Weiss, B., Jacquemin-Sablon, A., Live, T. R., Fareed, G. C., and Richardson, C. C. 1968. Enzymatic breakage and joining of deoxyribonucleic acid. J. Biol. Chem. 243:4543-4555.

Wilson, G. G., and Murray, N. E. 1979. Molecular cloning of the DNA ligase gene from bacteriophage T4. I. Characterization of the recombinants. J. Mol. Biol. 132:471–491.

EXERCISE 8

Ligation of Restriction Endonuclease Cleaved Chromosomal DNA Fragments and Plasmid Vector DNA

Materials

T4 DNA Ligase Buffer (10X):   200 mM Tris-HCl, pH 7.5, 100 mM $MgCl_2$, 100 mM dithiotreitol.

T4 DNA Ligase Storage and Dilution Buffer: 10 mM K_2HPO_4/ KH_2PO_4 pH 7.6, 50 mM KCl, 10 mM 2-mercaptoethanol, 0.5 mM NaN_3, 50% glycerol.

T4 DNA Ligase: 1.0 Weiss unit/ligation reaction.

ATP (10X):   5 mM.

Reaction Stop Mix: 5 M urea, 10% glycerol, 0.5% SDS, 0.025% xylene cyanole FF, 0.025% bromphenol blue WS.

Water (ddH_2O):   Sterile double distilled.

Note: One Weiss unit is the amount of T4 DNA ligase required to catalyze the conversion of 1 nmole of ^{32}PPi to $[\alpha/\beta\ ^{32}P\text{-ATP}]$ at 37°C in 20 minutes.

Protocol

DAY 1

1. Once it has been established that the plasmid and chromosomal DNA samples have been digested to completion, add Reaction 4–4 to Reaction 4–9 (*Pst*I digested pBR329 and *E. coli* DNA) and Reaction 4–7 to Reaction 4–8 (*Hind*III digested pBR329 and *E. coli* DNA). Dialyze these reaction mixtures against 1 liter of TEN buffer for at least 3 hours at room temperature. (See Appendix 2 for mini-dialysis procedure.)

2. After dialysis, set up the following ligation reactions in 1.5 ml microfuge tubes:

Reaction 6–1	μl	*Reaction 6–2*	μl
*Hind*III cut pBR329 + *Hind*III cut *E. coli* DNA	30	*Pst*I cut pBR329 DNA + *Pst*I cut *E. coli* DNA	30
1 unit T4 DNA ligase	1	1 unit T4 DNA ligase	1
10X T4 Ligase Buffer	10	10X T4 Ligase Buffer	10
10X ATP	10	10X ATP	10
ddH_2O	49	ddH_2O	49
	100		100

EXERCISE 8

LIGATION OF RESTRICTION ENDONUCLEASE CLEAVED CHROMOSOMAL DNA FRAGMENTS AND PLASMID VECTOR DNA

3. Incubate the reaction mixtures at 12.5°C for 12 to 18 hours. If a refrigerated water bath is not available, incubate the reaction in the refrigerator (4°C) until the next lab period (1 to 2 days).

4. At the end of the incubation period, remove about 0.2 μg (2μl) of the ligated DNA and add it to 8 μl of Reaction Stop Mix. Store the remaining ligation reactions at −20°C, until the extent of ligation can be determined visually by agarose gel electrophoresis. (The ligation can be considered complete when the plasmid vector DNA can no longer be visualized as a discrete band in the gel, Fig. 6.4d).

5. Terminate the ligation reactions by incubating at 65°C for 5 minutes. Store reactions at −20°C.

   Note: If a ligation reaction did not go to completion, thaw the reaction, add another 5 μl of ATP and 1 unit of T4 DNA ligase, and continue the incubation at 12.5°C for another 12 hours.

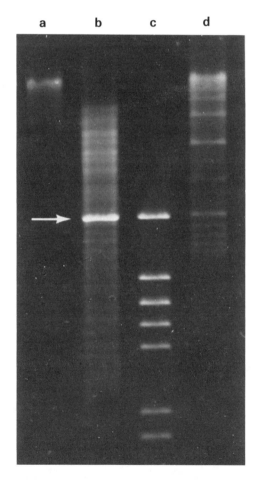

Figure 6.4
Agarose gel (1%) electrophoresis pattern of unligated and ligated DNA molecules: (a) Undigested *E. coli* chromosomal DNA. (b) *Pst*I cleaved *E. coli* chromosomal DNA and pBR329 plasmid DNA (indicated by arrow). (c) *Hin*dIII cleaved φ29 DNA as molecular weight standards. (d) Ligated products of Lane b.

CHAPTER 7

Bacterial Transformation

Unless introduced into a biological host, the recombinant DNA molecules formed by *in vitro* ligation represent mere chemical curiosities and they will degrade in a matter of hours unless special precautions are taken. Therefore, without the development of methods for introducing DNA molecules into common laboratory microorganisms such as *E. coli* and yeast, the power of the recombinant DNA approach would not have been fully realized. In retrospect, it should be pointed out that microbes were exchanging genetic material long before researchers began their analysis of gene structure and function. Exchange of genetic information between bacteria is generally accomplished by conjugation, transduction, and transformation.

Conjugation

The first report concerning the ability of *E. coli* to exchange genetic information was made in 1946 by Lederberg and Tatum. When grown together, two different auxotrophic strains were found to give rise to prototrophs. This form of genetic exchange requires the physical contact of the bacteria, and it is termed *conjugation*. Two bacterial phenotypes have been defined with respect to conjugation: F^+, indicating the ability to act as a donor of genetic information; and F^-, indicating the ability to act as a recipient of genetic material. Bacterial conjugation has two consequences: the transfer of genetic information from the donor to the recipient, and the conversion of the recipient from F^- to the F^+ phenotype. The physical basis of these events can be explained by the existence of the F (fertility) factor, an extrachromosomal element able to mobilize and transfer from one strain to another. In the process, regions of the bacterial chromosome may also be mobilized and transferred. Recombinational events in the recipient strain integrate the mobilized DNA into the chromosome and result in the formation of stable recombinants.

CHAPTER 7
BACTERIAL TRANSFORMATION

Mobilization of chromosomal genes by the F factor occurs at a low frequency and in a random manner. The Hfr mating phenotype was isolated on the basis of its ability to promote a high frequency of chromosomal transfer and recombination. The Hfr strain is able to function only as an efficient donor, not as a recipient, but the Hfr phenotype is not transferred to the recipient as is the case with the transfer of F^+. The Hfr strain can lose the ability to promote efficient genetic transfer, reverting to the typical F^+ phenotype. Unlike transfer from an F^+ strain, transfer from the Hfr strain is nonrandom, and certain groups of linked markers are transferred at high frequency relative to other markers. The physical basis of these properties of Hfr strains appears to be a result of the integration of the F factor into the bacterial chromosome. The markers that are preferentially transferred by each Hfr strain and the order in which they are transferred is determined by the location and orientation of the integrated F factor. Transfer of the chromosome initiates at the site where F is integrated, and transfer proceeds in such a manner that the fertility genes are the last region transferred. There appear to be a limited number of sites in the bacterial chromosome where the sex factor can be integrated. The specific nature and high frequency of transfer have made Hfr strains valuable in the construction of the chromosomal genetic map of *E. coli* (fig. 7.1).

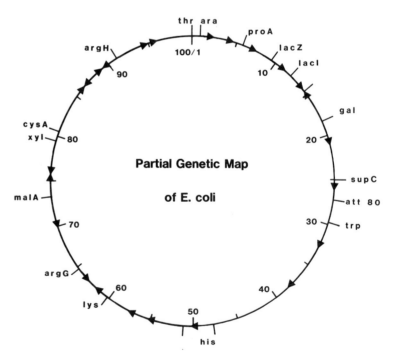

Figure 7.1
A summary of the F-factor integration sites in the chromosome of several Hfr *E. coli* strains. Each point of insertion is marked by an arrowhead pointing in the direction of chromosome transfer during conjugation. The gene behind the arrow is the first gene to be transferred by the Hfr.

Transduction

Although the various forms of the sex factor F are involved in the transfer of chromosomal genes between bacteria, transfer can also be mediated by bacteriophage. Temperate bacteriophage are capable of lysogeny, or integration into the bacterial chromosome. During induction of the lysogenized phage, recombinational events can produce a circular DNA molecule composed of a portion of the phage DNA linked to a portion of the chromosomal DNA adjacent to the site of phage lysogeny. These defective phage particles can be packaged in bacteriophage particles and released from the infected cell. Injection of the defective phage genome into a recipient cell introduces the bacterial genes; lysogenization of the defective phage genome results in the transfer of a segment of DNA derived from the previous attachment site of the prophage. This mode of genetic exchange has been termed *restricted transduction*.

A number of phage are capable of mediating a process called *generalized transduction*. In generalized transduction, each chromosomal marker has a similar chance of being transferred as a transducing particle. The transducing particles consist of a fragment of bacterial DNA packaged in a phage coat and contain no phage DNA. A bacterial lysate prepared with a generalized transducing phage contains both viable phage particles and particles of packaged chromosomal DNA. Infection of a recipient strain of bacteria with this lysate will introduce bacterial DNA into a fraction of the cell population. Recombinational events can result in the integration of this DNA and a change in the phenotype of the recipient bacterium.

Transformation

Transformation has been defined as the heritable change in the properties of one bacterial strain by the DNA of another. Transformation is a naturally occurring phenomenon that is not restricted to the laboratory. The release of DNA fragments that accompanies the death and lysis of a cell provides a source of genetic material for recombination during the growth of microorganisms in the natural environment. Since its discovery in *Pneumococcus*, genetic transformation has been observed in a variety of bacteria, including *Streptococcus pneumoniae, Bacillus subtilis, B. stearothermophilus, Haemophilus influenzae*, and *Acinetobacter*. In these naturally occurring transformation systems, the bacteria generally enter a stage of growth known as competence, during which they are capable of taking up exogenous DNA. Competent cells are able to bind nucleic acid in a form that is resistant to the action of nucleases. *Haemophilus* and *Bacillus* have been found to contain DNA binding proteins whose function appears to be necessary for this initial stage of transformation. Following uptake, the DNA may undergo a period known as the eclipse phase, during which it remains resistant to the action of nuclease but its genetic information is not expressed. This is the time during which the DNA is thought to be undergoing recombination with homologous sequences on the bacterial chromosome. Following the completion of the recombinational events, the genetic information on the incorporated DNA can be expressed, resulting in the appearance of the functionally transformed cell (fig. 7.2).

CHAPTER 7
BACTERIAL TRANSFORMATION

Early attempts to transform *E. coli* were unsuccessful, suggesting that this bacterium does not possess a natural mechanism for transformation. It was subsequently found that competence could be artificially induced by exposing cells to calcium chloride prior to the addition of DNA. The procedure originally reported by Mandel and Higa involved exposure of the growing bacteria to a hypotonic solution of calcium chloride at 0°C, causing the cell to swell (spheroplast formation). DNA added to the transformation mixture forms a DNase resistant complex of hydroxyl-calcium phosphate that adheres to the cell surface. This complex can be taken up by the cell during a brief 42°C heat pulse (fig. 7.3). After a few hours of growth in rich medium to allow the spheroplast to recover and the transformed genes to be expressed, transformants can be isolated by plating on selective medium. For example, cells transformed by pBR322 DNA can be selected on medium containing the antibiotics ampicillin and tetracycline. With minor modifications, this basic transformation protocol has been made to work for a variety of bacteria that do not take up exogenous DNA normally. It has recently been suggested that magnesium ions may play an important role in the stability of the DNA during uptake, and many transformation methods now include $MgCl_2$ during the treatment of the bacteria. Specialized transformation procedures have been developed for use with particular strains; for example, the biosafety containment strain $\chi1776$. As this strain is easily

DNA Mediated Transformation

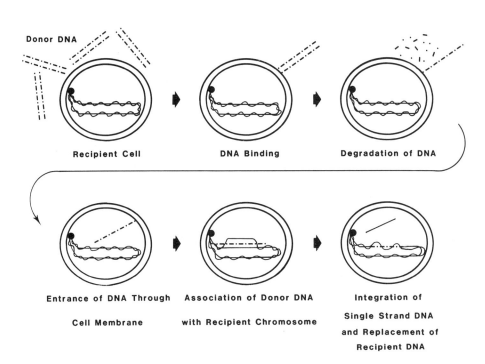

Figure 7.2

lysed, special precautions are necessary to obtain efficient transformation (see Appendix 2).

An alternate method for introducing foreign DNA into cells has been developed by taking advantage of the ability of a bacteriophage particle to efficiently inject DNA into a bacterial cell. Empty bacteriophage λ head precursor particles are filled with foreign DNA and converted to functional phage heads. These particles are then able to inject their contents through λ receptor sites on the recipient bacteria with a high efficiency. This method can be used with bacteriophage λ vectors or with the specialized cosmid vectors to introduce recombinant DNA molecules into bacteria. The principles of *in vitro* packaging and their use in cloning of DNA fragments have been described in detail in the references by Collins, 1979; Hohn, 1979; and Enquist and Sternberg, 1979.

CHAPTER 7
BACTERIAL TRANSFORMATION

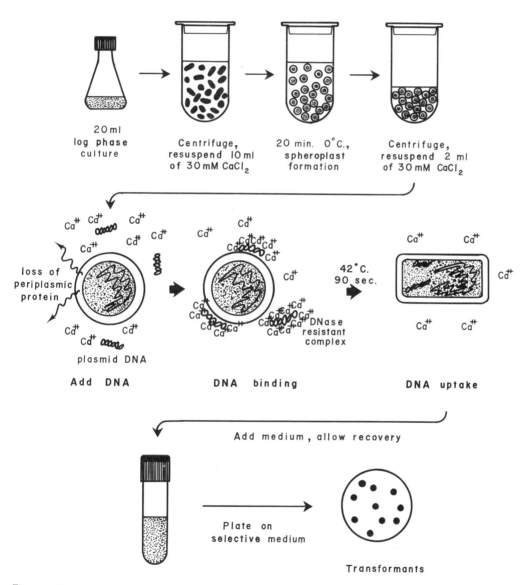

Figure 7.3
A diagrammatic representation of the CaCl$_2$-induced transformation procedure for *E. coli*.

CHAPTER 7
BACTERIAL TRANSFORMATION

The introduction of DNA into a cell is one aspect affecting the overall efficiency of transformation. A second important aspect is the stability of the DNA once it has entered the cell. A circular plasmid molecule has been found to have a 10 to 100 fold higher efficiency of transformation of *E. coli* than the linear form of the same plasmid molecule. A circular molecule entering the cell is resistant to the action of exonucleases present in the cell, whereas the linear molecule is susceptible to degradation of the free ends. Transformation with a linear molecule involves not only the entry of the DNA into the cell, but also the conversion of the linear molecule to a circular form to allow replication of the plasmid. This circularization can be prevented by alteration of the ends of the molecule. Successful circularization and replication of the linear plasmid DNA involves a competition between the action of DNA ligase forming the circular molecule and exonucleases degrading the linear molecule.

Bacteria also possess a restriction-modification system. The host restriction endonucleases will cleave DNA that has not been previously modified by a modification enzyme of the same strain. If plasmid DNA purified from *E. coli* C600 is used to transform *E. coli* K-12, the K-12 host restriction system will cleave the unmodified C600 DNA as it enters the cell and greatly reduce the efficiency of transformation. In contrast, DNA purified from a K-12 strain will have been modified by the host modification system, so that transformation of a different strain of *E. coli* K-12 will occur at a much higher transformation frequency. When DNA from an organism other than *E. coli* is used in the transformation process, the DNA will not have the proper modification pattern to prevent degradation by the host restriction system. To overcome this difficulty, the strains of *E. coli* routinely used for the maintenance of recombinant DNA molecules are generally deficient in the host restriction system ($hsdR^-$) and may also be deficient in the host modification system ($hsdM^-$). The use of the $hsdR^-$ phenotype prevents the restriction of the incoming DNA and greatly increases the efficiency of transformation with nonhomologous DNA. Although the presence of a functional modification system in the absence of a functional restriction system has little effect on the transformation efficiency with nonhomologous DNA, plasmids purified from a r^-m^+ strain can be subsequently transferred into a strain with the same restriction phenotype. For example, DNA purified from a strain of *E. coli* K-12 that is $r^-_k m^+_k$ can be used to efficiently transform a strain that is $r^+_k m^+_k$. Finally, in addition to those factors mentioned above, transformation frequencies can also be affected by: (1) the efficiency of spheroplast formation, (2) cell viability after $CaCl_2$-treatment and storage, and (3) DNA concentration, conformation, and purity.

Procedures similar to those used to transform *E. coli* have also been developed to allow the transformation of a variety of eukaryotic cells. The yeast *Saccharomyces cerevisiae* can be converted to spheroplasts with enzymes that digest the cell wall (e.g., glusulase, lyticase, and zymolyase). When the spheroplasts are exposed to DNA in the presence of polyethylene glycol and calcium chloride, the DNA is taken up as a result of spheroplast fusion. After the fused spheroplasts have had time to recover physiologically and regenerate the cell wall, transformants can be identified by plating on selective media. Just as calcium chloride has been used to induce competence in *E. coli,* lithium acetate and cesium chloride can be used (instead of enzymes) to induce competence in yeast (see Appendix 2). Using these chemical treatments, yeast can be transformed more efficiently and in shorter time than with those procedures that employ exogenous enzymes. In addition to yeast, mammalian tissue culture cells can be transformed using calcium chloride and DEAE-dextran treatments to stimulate DNA uptake. A number of transformation procedures for plant cells or plant protoplasts are also currently under development. While transformation procedures may differ in the kinds and concentrations of reagents used, they all share at least two common features. First, the cell membrane of the recipient host is destabilized by treatment with divalent cations (Ca^{++} or Mg^{++}) or by polyanions such as DEAE-dextran. Second, the transforming DNA is precipitated into a DNase-resistant calcium phosphate complex that stimulates DNA uptake. It appears that these two processes can occur separately and are essential for efficient transformation.

References

Alphen, W. V., Selm, N. V., and Lugtenberg, B. 1978. Pores in the outer membrane of *E. coli* K12—involvement of proteins b and e in the functioning of pores for nucleotides. Molec. gen. Genet. 159:75–83.

Altherr, M. R., Quinn, L. A., Kado, C. I., and Rodriguez, R. L. 1982. Transformation and storage of competent yeast cells. In Genetic Engineering in Eukaryotes, Plenum Press, New York.

Cohen, S. N., Chang, A. C. Y., and Hsu, L. 1972. Non-chromosomal antibiotic resistance in bacteria: Genetic transformation of *Escherichia coli* by R-factor DNA. Proc. Natl. Acad. Sci. USA 69:2110–2114.

Collins, J. 1979. *Escherichia coli* plasmids packageable *in vitro* in λ bacteriophage particles. In Methods in Enzymology, R. Wu, ed., 68:309–326.

Cosloy, S. D., and Oishi, M. 1973. The nature of the transformation process in *E. coli* K12. Molec. gen. Genet. 124:1–10.

Dargert, M., and Ehrlich, S. D. 1979. Prolonged incubation in calcium chloride improves the competence of *Escherichia coli* cells. Gene 6(1):23–28.

Eisenstadt, E., Lange, R., and Willecke, K. 1975. Competent *Bacillus subtilis* cultures synthesize a denatured DNA binding activity. Proc. Natl. Acad. Sci. USA 72:323–327.

Enquist, L., and Sternberg, N. 1979. *In vitro* packaging of λ*Dam* vectors and their use in cloning DNA fragments. In Methods in Enzymology, R. Wu, ed., 68:281–298.

Graham, F. L., and Van der Eb, A. J. 1973. A new technique for the assay of infectivity of Human Adenovirus 5 DNA. Virology 52:456–467.

Hayes, W. 1976. The genetics of bacteria and their viruses: Studies in basic genetics and molecular biology. John Wiley and Sons, New York.

Hinnen, A., Hicks, J. B., and Fink, G. R. 1978. Transformation of yeast. Proc. Natl. Acad. Sci. USA 75:1929–1933.

Hohn, B. 1979. *In vitro* packaging of λ and cosmid DNA. In Methods in Enzymology, R. Wu, ed., 68:299–309.

**CHAPTER 7
BACTERIAL
TRANSFORMATION**

Lederberg, J., and Tatum, E. L. 1946. Novel phenotypes in mixed cultures of biochemical mutants of bacteria. Cold Spring Harbor Symp. Quant. Biol. 11:113–114.

Mandel, M., and Higa, A. 1970. Calcium-dependent bacteriophage DNA infection. J. Mol. Biol. 53:159–162.

Mercer, A. A., and Loutit, J. S. 1979. Transformation and transfection of *Pseudomonas aeruginosa:* Effects of metal ions. J. Bacteriol. 140: 37–42.

Mertz, J., and Berg, P. 1974. Defective Simian Virus 40 genomes: Isolation and growth of individual clones. Virology 62:112–124.

Morrison, D. A. 1979. Transformation and preservation of competent bacterial cells by freezing. In Methods in Enzymology, R. Wu, ed., 68:326–330.

Notani, N. K., and Setlow, J. K. 1974. Mechanism of bacterial transformation and transfection. Proc. Nucl. Acids Res. Mol. Biol. 14:39–100.

Reid, J. D., Stoufer, S. D., and Ogrydziak, D. M. 1982. Efficient transformation of *Serratia marcescens* with pBR322 plasmid DNA. Gene 17:107–112.

Strike, P., Humphreys, G. O., and Roberts, R. J. 1979. Nature of transforming deoxyribonucleic acid in calcium-treated *Escherichia coli.* J. Bacteriol. 138:1033–1035.

Sutrina, S. L., and Scocca, J. J. 1979. *Haemophilus influenzae* periplasmic protein which binds deoxyribonucleic acid: Properties and possible participation in genetic transformation. J. Bacteriol. 139:1021–1027.

Wigler, M., Pellices, A., Silverstein, S., Axel, R., Urlaub, G., and Chasin, L. 1979. DNA-mediated transfer of the adenine phosphoribosyltransferase locus into mammalian cells. Proc. Natl. Acad. Sci. 76:1373–1376.

Wigler, M., Silverstein, S., Lee, L., Pellices, A., Cheng, Y., and Axel, R. 1977. Transfer of purified Herpes virus thymidine kinase gene to cultured mouse cells. Cell 11:223–232.

Zaenen, I., van Larebeke, N., Tuechy, H., van Montagu, M., and Schell, J. 1974. Supercoiled circular DNA in crown-gall inducing *Agrobacterium* strains. J. Mol. Biol. 86:109–127.

EXERCISE 9

Preparation and Transformation of Competent Cells of E. coli K12

Materials

Luria Broth:
  2, 20 ml liquid cultures in 250 ml erlenmeyer flasks.
  2 agar (1.5%) plates.
  4 agar (1.5%) plates containing 20 µg/ml ampicillin (Ap).

M9 Minimal Agar (Tc-M9 min.-ara): 10 plates supplemented with 0.2% arabinose, 166 µg/ml proline, 41 µg/ml leucine, 0.2 µg/ml thiamine hydrochloride (B_1) and 10 µg/ml tetracycline

M9 Minimal Agar (Ap-M9 min.-hisol.): 10 plates supplemented with 0.2% glucose, 166 µg/ml proline, 0.2 µg/ml B_1, 20 µg/ml histidinol, and 20 µg/ml ampicillin.

Storage Mixture: 30 mM $CaCl_2$, 15% glycerol, sterile.

λ Diluent (λ dil): Sterile, 10 mM Tris-HCl pH 7.6, 50 mM NaCl, and 0.01% gelatin.

0.3M $CaCl_2$: Sterile.

30 mM $CaCl_2$: Sterile.

Ice.

Note: Using a 1:100 inoculum of an overnight culture, start the 20 ml LB cultures of *E. coli* LA6 and BE42 cells at least 3 hours before the exercise.

Protocol

DAY 1

1. Harvest the cells from two, 20 ml log-phase LB cultures of LA6 and BE42 (approximately $2-5 \times 10^8$ cells/ml) by centrifugation. Pour the cultures into sterile, 25 ml screw-cap Corex tubes (or sterile 50 ml polypropylene tubes with caps) and spin down the cells at 10 krpm for 5 minutes at 4°C. All steps in this protocol must be performed aseptically.

EXERCISE 9
PREPARATION AND TRANSFORMATION OF COMPETENT CELLS OF *E. COLI* K12

2. Pour off the liquid medium and resuspend the cell pellet in 1/2 volume (10 ml) of cold 30 mM $CaCl_2$, by vortexing vigorously. Let the cell suspension sit on ice for 20 minutes. This treatment promotes spheroplast formation. Since the cells will become swollen and fragile, they must be handled gently in all subsequent steps.

3. Harvest the cells by centrifugation at 5 krpm for 5 minutes at 4°C. Pour off the supernatant and resuspend the cell pellet in 1/10 volume (2 ml) of cold 30 mM $CaCl_2$, 15% glycerol. Hold the suspension on ice until it is ready to be used or stored. (The cell pellet can be gently resuspended by pipeting the solution several times with a 10 ml serological pipet.)

4. To store the cells for future use, label 20 sterile microfuge tubes (10 for each strain) with your initials, date, and strain designation. With the microfuge tubes sitting on ice, dispense 0.2 ml of the $CaCl_2$-treated cells into each tube. Cap the tubes and place at −70°C. Stored in this fashion, the competent cells will remain viable for 3 months.

DAY 2

1. To test the competence of the LA6 and BE42 cells, add 2 µl of purified pBR329 DNA (0.5 µg/ml) to 1 microfuge tube containing LA6 cells and 1 microfuge tube containing BE42 cells. As a control, add 2 µl of 30 mM $CaCl_2$ to two additional microfuge tubes containing LA6 and BE42 cells. Gently mix the contents of the 4 tubes and let them stand on ice for 1 hour.

2. Place the microfuge tubes containing the pBR329 DNA at 42°C for 90 seconds, then place them on ice for 5 minutes. For the BE42 cells ($r^+_k m^+_k$), extend the heat pulse to 6 minutes and let stand on ice for 5 minutes. This helps to inactivate the host restriction system.

3. Place the contents of each microfuge tube into a sterile test tube. Add 3 to 4 ml of LB and incubate at 37°C for 3 hours with gentle shaking. Although portions of the transformation mixture can be plated directly on selective antibiotic medium, the frequency of transformation is usually lower than those mixtures that have had time to recover and phenotypically express plasmid encoded genes.

4. Spread 0.1 ml of each transformation mixture on separate LB-Ap plates. As a second control (for viability), make a 10^{-4} dilution (0.05 ml in 5 ml of LB, twice) of the two DNA-containing transformation mixtures and spread 0.1 ml from the final dilutions on separate LB plates. Label the backs of all plates with your initials, date, and strain designation, and incubate (face down) at 37°C overnight.

DAY 3

Calculate the frequency of transformation for the two strains of *E. coli*, in terms of the number of transformants (Ap^r colonies) per viable cell (colonies on LB plates) per microgram DNA. If the $CaCl_2$-treated cells prepared on Day 1 are found to be competent (i.e., ~10^{-4} transformants/μg DNA), begin preparing ligation reactions 6-1 and 6-2 (Chapter 6) for transformation.

EXERCISE 9

PREPARATION AND
TRANSFORMATION OF
COMPETENT CELLS OF E. COLI K12

1. Spin Reactions 6-1 and 6-2 in the microcentrifuge for 2 minutes.

2. Pipet off the upper 90 μl from each tube, taking care not to touch the bottom of the microfuge tube with the micropipet.

3. Transfer the 90 μl (now free of contaminating microorganisms) to a sterile glass test tube (10 mm × 100 mm). Add 10 μl of sterile 0.3 M CaCl$_2$; mix and put on ice. Notice the formation of a granular white precipitate.

4. Thaw two tubes of competent LA6 and BE42 cells by letting them stand on ice for 15 to 20 minutes. Add the BE42 cells to the test tube containing Reaction 6-1, and the LA6 cells to the tube containing ligation reaction 6-2. Mix and let stand on ice for 1 hour.

5. Heat the tube containing the LA6 to 42°C for 90 seconds, and the BE42 cells for 6 minutes at 42°C. Put both tubes on ice for 5 minutes.

6. Add 3 ml of LB to each test tube and incubate at 37°C for at least 3 hours with gentle shaking. For convenience, cultures can also be incubated overnight.

7. Before spreading the transformation mixtures on selective media, wash the cells free of LB by centrifugation at 5 krpm for 5 minutes at 4°C. Resuspend the cell pellets in 3 ml of λ dil. Spin the cells down again and resuspend the cell pellet in 1/2 volume (1.5 ml) of λ dil.

8. Spread 0.1 ml of the LA6/Reaction 6-2 transformation mixture onto each of 5 Tc-M9 min.-ara plates.

9. Spread 0.1 ml of BE42/Reaction 6-1 transformation mixture onto each of 5 Ap-M9 min.-hisol. plates. Label all plates and incubate as described in Step 4, Day 2.

10. As a control, wash (in λ dil) and plate 0.1 ml of the pBR329/LA6 and pBR329/BE42 transformation mixtures (Step 4, Day 2) on a Tc-M9 min.-ara and Ap-M9 min.-hisol. plate, respectively. Label and incubate plates as previously described.

11. Add 3 ml of LB to each of the remaining transformation cultures (1 ml λ dil + 3 ml LB). Store at 4°C.

Note: It may be necessary to incubate the minimal plate for 2 to 4 days before colonies are visible. Record the number of Ap^r and Tc^r transformants obtained from the two sets of plates. If less than 10 (for each of the set of 5 plates), spread more of the transformation mixture as described in Steps 7, 8, and 9, Day 3.

CHAPTER **8**

Identification and Characterization of Recombinant Transformants

After ligation of DNA fragments to a vector and transformation of the host with the ligated products, it is necessary to distinguish recombinant DNA molecules from nonrecombinant molecules. Even with cloning strategies that require the formation of recombinant molecules for successful transformation (e.g., alkaline phosphatase treatment and A/T or G/C tailing) it is still necessary to distinguish the desired recombinant molecule from those generated by aberrant recombinational events. The identification of recombinant plasmids can be accomplished using direct or indirect means. Indirect methods frequently involve scoring transformants for a new or altered phenotype conferred by the recombinant plasmid. For example, the mouse dihydrofolate reductase gene cloned into the *Pst*I site of pBR322 confers a Tc^r, Ap^s, Tp^r (trimethoprim resistance) phenotype to the bacterial host. More direct methods may involve: (1) the physical isolation and restriction enzyme characterization of plasmid DNA, (2) the *in situ* or *in vitro* hybridization of plasmid DNA to complementary radiolabeled probe, and (3) radioimmune detection of cross-reacting material expressed by the recombinant plasmids. Since the desired recombinant transformant may represent only a minute fraction of the transformed population, it is quite common to use a combination of direct and indirect methods to achieve positive identification in the shortest period of time and with the least amount of effort.

Phenotypic Characterization

Following the transformation of bacteria with a ligated DNA sample, aliquots of the transformation are plated on selective medium in order to detect the growth of transformed cells. The appearance of colonies does not guarantee that the desired transformants have been obtained. The initial transformants must be characterized phenotypically and plasmid DNA must be extracted from each transformant to confirm the presence of recombinant DNA molecules. Finally, the original auxotrophic strain must be transformed again to show that the phenotypic marker selected coincides with plasmid markers.

CHAPTER 8
IDENTIFICATION AND CHARACTERIZATION OF RECOMBINANT TRANSFORMANTS

As discussed in Chapter 2, bacterial colonies that appear on a selective plate are not necessarily derived from the growth of a single cell. In a typical experiment, 10^8 competent cells will be exposed to a DNA sample. If the efficiency of transformation is 10^{-6}, then approximately 100 of these cells will be transformed. If the entire transformation mix is plated on selective medium, 100 colonies will arise out of a background of 9.9999×10^7 cells that have not been transformed. Under many selective conditions, cells that are adjacent to a transformant will begin to grow once the growth of the transformant has been established, and the resulting colonies may be a mixture of transformed and nontransformed cells. This phenomenon is known as *crossfeeding*. An important step in characterizing a transformant involves the isolation of a pure colony derived from the growth of a single cell. This can be easily done by streaking the transformants on selective medium to obtain single-colony isolates as described in Exercise 1.

Many vectors are available that allow the detection of recombinant molecules by the insertional inactivation of one of the phenotypic markers of the vector. Transformants are obtained initially by selecting for a vector encoded marker that has not been affected by the cloning process. Among the transformants, recombinants can be identified by scoring for markers affected by the cloning process. For example, insertion of DNA fragments in the *Pst*I site of pBR322 inactivates the ampicillin resistance mechanism but does not affect tetracycline resistance. Tetracycline resistant transformants can be scored for ampicillin sensitivity to detect the recombinant plasmids. Using a numbered template positioned under a petri dish containing selective medium, several hundred transformants can be examined in an orderly fashion. A minimum of two plates are generally used for this purpose: a master plate of the same type as that used to obtain transformants, and a test plate that can distinguish recombinant from nonrecombinant transformants. The phenotype of each transformant is determined or scored in terms of its growth on the test plate. The master plate serves to verify the original phenotype. Note that when the test plate involves a negative selection (i.e., no cell growth) as a means of detecting recombinant transformants, a master plate serves as a source of the desired transformant for future studies. This procedure is illustrated in Fig. 8.1.

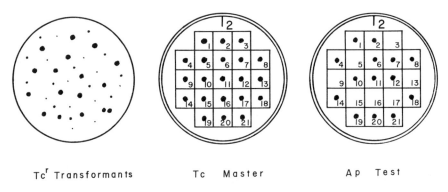

Tc^r Transformants   Tc Master   Ap Test

Figure 8.1
Patching method for screening tetracycline resistance (Tc^r) for ampicillin sensitive recombinant transformants.

CHAPTER 8

IDENTIFICATION AND CHARACTERIZATION OF RECOMBINANT TRANSFORMANTS

When the phenotype of large numbers of transformants must be examined to detect the desired recombinants, the technique of replica plating may be employed. In replica plating, a sterile cloth (usually velvet) or filter paper is applied to the surface of a plate containing bacterial colonies and carefully removed, retaining an imprint of the colonies. Master and test plates are then carefully pressed on this surface and removed. This transfers a sufficient number of cells to each plate so that, following incubation, colonies will appear on the surface of the replica plates. The composition of the master plate should be the same as the original plate, and growth on the test plates will indicate the phenotype of the colonies. This process is illustrated in Fig. 8.2. While replica plating allows the screening of many colonies in a relatively rapid manner, this technique requires some practice and in many situations may provide no great advantage over the testing method previously described.

Once colonies of the proper phenotype have been identified and purified, it becomes necessary to establish a causal relationship between the new or altered phenotype and the presence of a recombinant plasmid. This can be done by purifying sufficient amounts of plasmid DNA to allow transformation of competent cells and construction of a physical restriction cleavage map of the plasmid. A variety of rapid miniscreens are available that allow the purification of plasmid DNA from a small (1–10 ml) bacterial culture (see Chapter 3). Any one of these methods will provide sufficient DNA to accomplish the two objectives above.

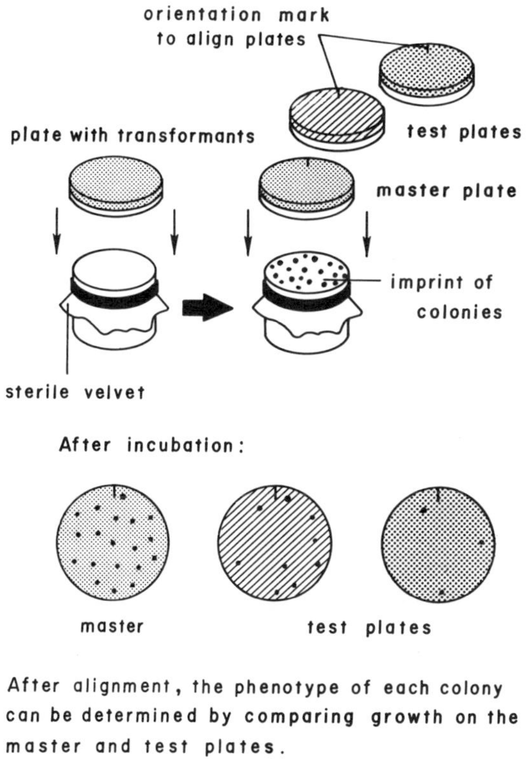

Figure 8.2
Screening recombinant transformants using the replica plating method.

Physical Characterization: Plasmid Isolation and Transformation

The isolation of a recombinant plasmid from a transformant of the desired phenotype is not unequivocal evidence that the gene of interest has been successfully cloned. To prove this, it is necessary to isolate plasmid DNA from the recombinant transformant and retransform the original host strain. The transformation culture should then be plated on the same selective media. At this point, if every new transformant has the same phenotype as the transformant from which the plasmid DNA was purified, it is probable that the recombinant plasmid is responsible for that selected phenotype.

The appearance of more than one phenotype after the retransformation step may be explained in several ways. The plasmid in the original transformant may be unstable and recombine or delete various regions to generate the variation in phenotype; the original transformant may have contained more than one recombinant plasmid; or a spontaneous chromosomal mutation may have arisen in the original transformant to give the observed phenotype. As an example of this retransformation step, consider two $Ap^r\ Tc^s\ gal^+$ transformants obtained when a gal^- strain was transformed with pBR322 carrying a piece of chromosomal DNA purified from a gal^+ strain and inserted at the SalI site. These two transformants would appear to contain the gal genes cloned in the Tc^r gene of pBR322. After purifying plasmid DNA from these transformants and retransformation of a gal^- strain, the DNA purified from the first transformant gives rise to transformants that are $Ap^r\ Tc^s\ gal^+$, while the DNA from the second transformant gives rise to transformants that are $Ap^r\ Tc^s\ gal^-$. The recombinant plasmid present in the first transformant does complement the gal^- phenotype and is likely to contain the desired DNA fragment. However, the phenotype of the second transformant is probably the result of the presence of a recombinant plasmid that confers the $Ap^r\ Tc^s$ phenotype accompanied by reversion of the gal^- phenotype of the host strain to gal^+. Alternatively, the competent cells used in the original transformation may have been contaminated with unrelated gal^+ bacteria. Although the retransformation step is not direct proof that the desired gene(s) is contained on the recombinant plasmid, it does demonstrate that the desired phenotype is associated with the presence of that plasmid.

Physical Characterization: Restriction Enzyme Analysis

The purification of a small amount of plasmid DNA also allows the construction of a restriction cleavage map of the plasmid. This is accomplished by digesting samples of the plasmid DNA with a variety of restriction endonucleases and examining the DNA fragments on agarose or polyacrylamide gels. The sizes of the fragments are determined by comparison to DNA fragments of known size (see Chapter 5). In order to construct a restriction map, it is necessary to orient the position of the cleavage sites on the fragment relative to known sites on the vector. This can be accomplished by the use of double digestions—the successive digestion of a DNA sample with two or more enzymes. The changes in fragment size that are the result of digestion with a second enzyme indicate the distances between the various cleavage sites. By comparing the possible orientations of the cleavage sites relative to each other with the fragment sizes generated by double digestion, the correct orientation of the sites can be determined. This is illustrated in Fig. 8.3, where a restriction map of the plasmid pWAL1 has been constructed. This plasmid was constructed by the insertion of a 4.1 megadalton fragment of walrus DNA into the EcoRI site of pBR322. Examination of the EcoRI and SalI digestion patterns reveals the presence of a single SalI site in the cloned fragment. There are two possible orientations of this site relative to the SalI site of pBR322, as shown in Fig. 8.3. Comparison of the fragment sizes predicted by these two orientations with the fragment sizes generated by digestion of pWAL1 with both EcoRI and SalI allows the correct position of the SalI site in the cloned fragment to be determined.

The construction of a restriction map often involves the positioning of two or more cleavage sites that are contained within a restriction fragment generated by another enzyme. This can be accomplished by digesting the DNA fragment with the second enzyme under conditions that generate partial digestion products. Digestion of the DNA with a limiting

CHAPTER 8

IDENTIFICATION AND CHARACTERIZATION OF RECOMBINANT TRANSFORMANTS

Digestion results for pBR322 and pWAL1.

pBR322		pWAL1			
SalI or EcoRI	SalI + EcoRI	EcoRI	SalI	SalI + EcoRI	Fragment Identity
		4.1	4.2		cloned fragment
2.7		2.7			pBR322 linear
	2.3		2.6	2.3	EcoRI-SalI pBR322
				2.2	EcoRI-SalI
				1.9	EcoRI-SalI
	.4			.4	EcoRI-SalI pBR322

Possibilities for generating the SalI fragments of pWAL1:

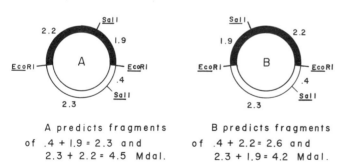

A predicts fragments of .4 + 1.9 = 2.3 and 2.3 + 2.2 = 4.5 Mdal.

B predicts fragments of .4 + 2.2 = 2.6 and 2.3 + 1.9 = 4.2 Mdal.

Digestion results indicate B is the correct map.

Figure 8.3
Restriction enzyme analysis of the plasmid pWAL1 using single and double digestions with EcoRI and SalI. Restriction fragment sizes are given in terms of megadaltons (Mdal). The solid portion of the circular plasmid map represents walrus DNA, the open portion, pBR322.

CHAPTER 8

IDENTIFICATION AND CHARACTERIZATION OF RECOMBINANT TRANSFORMANTS

amount of enzyme or preparing a time course of the digestion are two ways to observe the progression of the reaction from undigested to completely digested fragments. In Fig. 8.4, a time course of the digestion of a 6.5 kb fragment that contains two EcoRI sites has been illustrated. After 20 minutes of digestion, the complete digestion products of 3.0, 2.0, and 1.5 kb are apparent, while at the shorter digestion times, partial digestion products of 5.0 and 3.5 kb can be seen. Because each of the partial digestion products must be the sum of two or more of the complete digestion products, the size of the partial products can be used to determine which fragments are adjacent to one another. With this information, the location of both EcoRI cleavage sites can be assigned.

The construction of a simple restriction map serves two basic purposes. It confirms that the plasmids contained in transformants are in fact recombinant molecules that contain insertions of DNA in the expected locations, and it provides a physical map for the further characterization and manipulation of the recombinant molecule. The extent to which mapping of a plasmid should progress is to some extent dependent on the nature of the fragment and the subsequent experiments planned. With the current availability of some 40 restriction enzymes with different

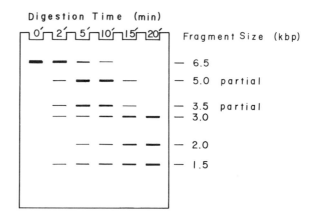

Figure 8.4
Restriction enzyme analysis using partial digestion with EcoRI. A diagrammatic representation of an ethidium bromide stained, UV irradiated agarose gel shows the EcoRI partial digestion products. Fragment sizes are given in terms of kilobase pairs (kbp). The intensities of the bands on the gel are intended to reflect the relative amounts of DNA in each band.

cleavage specificity, the use of every enzyme becomes time-consuming and prohibitively expensive. In general, restriction mapping of recombinant plasmids should begin with the enzymes that cleave at hexameric sites and proceed with enzymes that recognize pentamers and tetramers as the need indicates.

Physical Characterization: *In situ* (Colony) Hybridization

In many instances, a recombinant plasmid containing a desired DNA sequence must be identified in a population of bacteria containing many different recombinant molecules. When there is no genetic test for the presence of the desired recombinant, other screening methods must be employed. The technique of colony hybridization utilizes the principles of nucleic acid hybridization to detect recombinant molecules containing specific DNA sequences. Colonies or phage are transferred to a membrane support, such as a nitrocellulose filter, lysed, and the DNA denatured and bound to the filter *in situ*. The filters with the DNA imprint are then hybridized with a radiolabeled RNA or DNA probe specific for the desired DNA sequences. Following hybridization and removal of the unhybridized probe, autoradiography reveals which of the DNA imprints on the filter have homology with the probe. The colony from which the DNA was obtained contains the desired DNA sequences. Once identified, the colonies are recovered from a set of master plates made prior to immobilization and denaturation of the test colonies. Plasmid DNA is extracted from these colonies and characterized to verify the presence of the desired recombinant molecule. *In situ* hybridization is of particular value in the identification of specific recombinants in eukaryotic genomic clone banks, where hundreds of colonies or plaques must be examined to detect the desired recombinant. Modifications of the colony hybridization procedure have also been applied to the screening of yeast colonies and SV40-induced animal cell plaques.

Physical Characterization: Immunoassays

In some instances, it is possible to screen a gene bank for the presence of a specific recombinant plasmid by the use of immunoassays. These methods are dependent on the availability of an antibody directed against the protein of interest. If a cloned DNA fragment is faithfully transcribed and translated *in vivo*, the resulting protein product can be detected immunologically. A variety of detection methods have been utilized in various immunoassays, including the formation of precipitin bands in agar plates containing antibody, and the enzyme-linked immunosorbent assay (ELISA) for detection of protein (Kaplan et al., 1981). The most sensitive assays used for the immunologic screening of clone banks involve attachment of antibody or antibody fragments to a solid support, followed by exposure of the antibody-coated support to colonies that have been lysed on agar plates, as shown in Fig. 8.5. Sites of antigen binding can be detected by incubation with ^{125}I-labeled antibody against the desired antigen. Alternatively, sites of antigen binding can be detected by incubation with antibody against the antigen, followed by incubation

CHAPTER 8

IDENTIFICATION AND
CHARACTERIZATION OF
RECOMBINANT TRANSFORMANTS

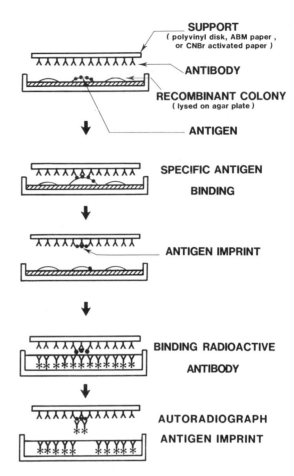

Figure 8.5
Immunodetection of antigen-expressing recombinant transformants using the two-site solid-phase radioimmune assay.

with ^{125}I-labeled A protein from *Staphylococcus aureus*, which will bind to the antibody present at sites of antigen-antibody interaction. Autoradiography reveals the colonies that produce the antigen recognized by the test antiserum. As with the colony hybridization technique, the desired recombinant transformant can be obtained from master plates and characterized further.

Localization of Genes on Recombinant Plasmids

Localizing a gene to a specific region of a cloned DNA fragment is an important step in the genetic analysis of a recombinant DNA molecule. The localization process not only identifies coding and regulatory regions, but also identifies intergenic or nonessential regions that can subsequently be removed from the cloned fragment to simplify the analysis. Two techniques of particular value in gene localization are subcloning and transposon mutagenesis. When combined with *in vivo* or *in vitro* transcription-translation systems, these methodologies can provide valuable information regarding the precise location and regulation of the gene of interest.

Subcloning

The process of subcloning involves the cloning of different portions of the original cloned DNA fragment in order to identify the smallest piece of DNA that contains the gene. The cloned genes may be functionally identified by complementation of a mutant host phenotype, or physically identified by hybridization of the subcloned DNA fragments with a radioactive probe (e.g., mRNA or cDNA). In the simplest form of subcloning, a DNA molecule known to contain a desired genetic region will be digested with restriction enzymes other than those used in the construction of the original recombinant molecule. The resulting fragments of the cloned genes can be ligated to a cloning vector and the products of ligation used to transform the host cell. The resulting recombinants are then screened for the desired genetic property. Each of the cloned fragments can be referred to as a subclone of the original recombinant molecule. For illustrative purposes, consider the recombinant plasmid pREC1 that consists of a 6 megadalton *Eco*RI fragment of *E. coli* DNA ligated to pBR322. Assume that pREC1 is able to complement the $genA^- \ genB^-$ phenotype of the strain of *E. coli* used as the recipient in the original transformation. The objective is to demonstrate that the genes *genA* and *genB* are located on pREC1.

As illustrated in Fig. 8.6, the construction of a restriction map of pREC1 reveals the presence of three internal *Sal*I sites, A, B, and C, in addition to the two *Eco*RI sites used in the construction of the recombinant. Digestion of pREC1 with *Eco*RI and *Sal*I will generate two *Sal*I fragments, $SalI_A\text{-}SalI_B$ and $SalI_B\text{-}SalI_C$, and two *Eco*RI-*Sal*I fragments, $EcoRI_A\text{-}SalI_A$ and $SalI_C\text{-}EcoRI_B$, each of which may contain *genA* or *genB*. The fragments $EcoRI_A\text{-}SalI_D$ and $SalI_D\text{-}EcoRI_B$ are both derived from pBR322 and do not contain either *genA* or *genB*. In order to determine by subcloning which of the fragments actually contains these genes, pREC1 can be digested with *Eco*RI and *Sal*I, then ligated with pBR322 that has been digested with *Sal*I or with *Sal*I and *Eco*RI. The *Sal*I site of pBR322 is in the gene for tetracycline resistance, and on the basis of their ampicillin resistant, tetracycline sensitive phenotype, the subclones pSUB1–4 can be readily identified among the transformant population and distinguished from one another by the use of plasmid miniscreens. Each of these subclones can then be used to transform a strain that is $genA^- \ genB^+$ and one that is $genA^+ \ genB^-$ and the phenotypic properties of the subclones determined. In this example, only pSUB1 is able to complement the $genA^-$ phenotype, and none of the subclones is able to complement the $genB^-$ phenotype. This observation suggests that the gene for *genA* is probably contained within pSUB1 and that the gene for *genB* is probably inactivated by cleavage of one of the *Sal*I sites.

In order to obtain a subclone that is functionally $genA^+ \ genB^+$, a partial digestion of pREC1 with a limiting amount of *Sal*I endonuclease can be used to remove one or more of the *Sal*I generated fragments. Following ligation under conditions that favor circularization of the digestion products, and transformation, selective conditions are used to identify transformants that are $genA^+ \ genB^+$. The resulting subclones, pPAR1–3, and their phenotypes are indicated in Fig. 8.7. Since $SalI_D$ is located within the tetracycline resistance gene of pBR322, the deletion of

CHAPTER 8

IDENTIFICATION AND CHARACTERIZATION OF RECOMBINANT TRANSFORMANTS

the fragment $SalI_D$-$SalI_C$ eliminates tetracycline resistance. The results indicate that the entire region from $SalI_B$ to $SalI_D$ can be deleted without affecting the expression of either *genA* or *genB*. When combined with the previously obtained information that fragment $SalI_A$-$SalI_B$ is sufficient for the *genA*$^+$ phenotype only, it can be concluded that the region from $EcoRI_A$ to $SalI_B$ is necessary for the *genA*$^+$ *genB*$^+$ phenotypes and that the region from $SalI_B$ to $EcoRI_B$ is not required. For further analysis of these genes, the plasmid pPAR1 may be of more use than pREC1, since 2 of the original 6 megadaltons of DNA has been removed.

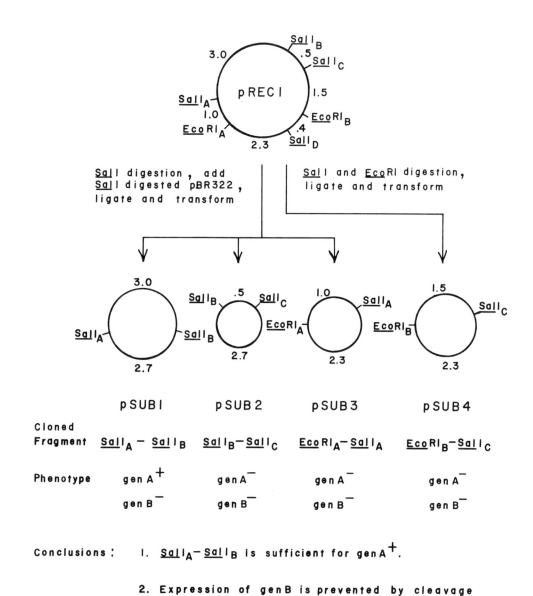

Figure 8.6
Gene localization by subcloning *Sal*I and *Sal*I/*Eco*RI restriction fragments of pREC1. Fragment and plasmid sizes are given in megadaltons (Mdal).

CHAPTER 8

IDENTIFICATION AND CHARACTERIZATION OF RECOMBINANT TRANSFORMANTS

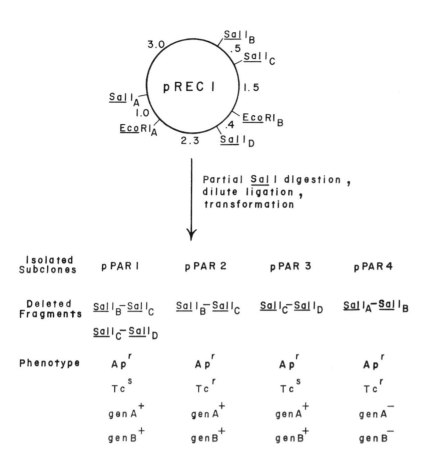

Figure 8.7
Gene localization using partial *Sal*I digestion of pREC1. Fragment sizes are given in megadaltons (Mdal).

Subcloning involves the deletion of various regions of a cloned DNA fragment in order to determine the regions involved in a genetic function. Similar information can be obtained by inserting DNA fragments into various regions of a cloned fragment. In the previously cited example, the locations of genes on pREC1 could have been determined by cloning an unrelated DNA fragment, such as a *Sal*I generated *D. melanogaster* DNA fragment, into each of the *Sal*I sites of pREC1. Insertion at $SalI_D$ would inactivate tetracycline resistance, insertion at $SalI_B$ and $SalI_C$ would have no detected phenotypic results, and insertion at $SalI_A$ would generate the phenotype $genA^+$ $genB^-$, indicating that $SalI_A$ is involved in the expression of *genB*. The use of restriction endonuclease cleavage sites as sites for insertional inactivation is convenient only when there are a limited number of the cleavage sites to be examined. Although each of the four *Sal*I sites in pREC1 can be readily examined by insertional inactivation, the presence of twenty-seven *Hae*II sites in pREC1 would make insertional inactivation of each of these sites laborious and impractical.

Insertional Inactivation Using Transposon Mutagenesis

113

CHAPTER 8

IDENTIFICATION AND CHARACTERIZATION OF RECOMBINANT TRANSFORMANTS

Fortunately, nature has provided a very efficient means for the insertion of DNA fragments. This approach utilizes a transposable genetic element (see Appendix 4). These are nonreplicating segments of DNA that are capable of inserting themselves into other DNA molecules. Many transposable elements carry genes that confer antibiotic resistance. Therefore, a transposition event can be detected by the expression of newly acquired antibiotic resistance genes on the recipient DNA molecule. The extensive characterization of a number of drug resistance transposons makes it possible to use transposon mutagenesis for the insertional inactivation and mapping of genes. To utilize the transposable element in this manner, a plasmid or phage containing a transposon is introduced into the bacteria containing the plasmid to be mutagenized. After allowing sufficient time for the element to transpose into the target plasmid, the plasmid DNA is purified and transformed into a recipient strain of bacteria. By selecting against a phenotypic marker present on the original mutagenizing vector, and for the drug resistance phenotype associated with the transposable element, transformants can be identified that contain plasmids in which the transposon has been inserted. Using restriction endonuclease cleavage maps of the transposon and the target plasmid, one can determine the various sites of transposon insertion. The phenotypic changes associated with each insertion can be determined by complementation, generating a map relating gene function with site of insertion.

As an example of gene mapping using Tn5 transposon mutagenesis, consider the plasmid pREC1 carrying the genes *gen*A and *gen*B. After pREC1 has been introduced into a strain containing the kanamycin resistance transposon Tn5, plasmid DNA is isolated and used to transform a kanamycin sensitive, ampicillin sensitive strain of *E. coli*. Transformants are then selected for resistance to both kanamycin and ampicillin. The position of each of the insertion events is determined by restriction mapping, and the phenotype associated with each insertion determined using complementation analysis in *gen*A$^-$ and *gen*B$^-$ strains of bacteria. The results, summarized in Fig. 8.8, indicate that in many regions of pREC1, insertions of Tn5 do not result in the loss of any phenotypic marker. In the region known to code for the tetracycline resistance gene,

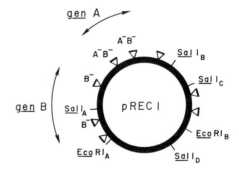

Figure 8.8
Gene localization using transposon mutagenesis of the plasmid pREC1. Triangles indicate sites of transposon *Tn*5 insertion. The locations of genA and genB, as determined by transposon mutagenesis, are indicated by the double-head arrows.

loss of tetracycline resistance is observed. Insertions in a second region result in the loss of either genB alone or both genA and genB. These results suggest that genA and genB are encoded on pREC1, as indicated in Fig. 8.8. In addition, the failure to observe the phenotype genA$^-$ genB$^+$ suggests that the expression of genA is necessary for the expression of genB and that the two genes may be transcribed as a single unit or polycistron with transcription initiation from genA.

A recombinant plasmid containing a cloned gene may prove to be a poor recipient for transposable elements. In such a case, it is possible to construct chromosomal mutants in which a transposon has been inserted into the gene of interest. Following the construction and characterization of this mutant, the recombinant plasmid can be introduced into the strain. Due to regions of homology between the recombinant plasmid and the chromosome, recombinational events can result in the transfer of a region of the chromosome containing the transposon to the recombinant plasmid. After the isolation of the altered recombinant plasmid, restriction mapping techniques can be used to map the site of insertion. Again, correlating phenotype with insertion site will yield information about the relative location of the genes.

Confirming Gene Location

The techniques described previously allow for the construction of a genetic map of a DNA molecule on the basis of subcloning and transposon mutagenesis. In order to verify that the inferred positions of the genes are correct and that the genes code for the correct gene products, it may be necessary to transcribe and, in the case of protein products, translate the DNA fragment to confirm genetic function. Three systems are commonly used to observe the protein products encoded by recombinant plasmids: *E. coli* minicells, *E. coli* maxicells, and *in vitro* transcription-translation systems. Unfortunately, all of these methods involve the use of radioisotopes and are sufficiently complicated to be beyond the scope of this laboratory manual. However, the importance of these techniques to the analysis of recombinant DNA molecules warrants a brief discussion of each method.

The minicell system involves the use of a minicell-producing strain of *E. coli*. Minicells are produced as a result of a mutation that causes the bacterial cell to bud-off a small inviable cell (fig. 8.9). Minicells normally contain no DNA, but if produced from a plasmid-containing strain, plasmid molecules can segregate into the forming minicells to produce miniature cells containing plasmid DNA instead of chromosomal DNA. Because of the size difference between minicells and viable cells, minicells can be purified from viable cells by methods such as velocity sedimentation (fig. 8.10). A purified preparation of plasmid-containing minicells is capable of RNA and protein synthesis for a short period of time after purification. When incubated in the presence of radioactive precursors, such as ^{35}S-methionine, the proteins synthesized by the purified minicells will be the result of plasmid-directed protein synthesis. The radioactive polypeptides can then be analyzed by SDS-polyacrylamide gel electrophoresis and autoradiography. Examination of the polypeptides syn-

CHAPTER 8

IDENTIFICATION AND
CHARACTERIZATION OF
RECOMBINANT TRANSFORMANTS

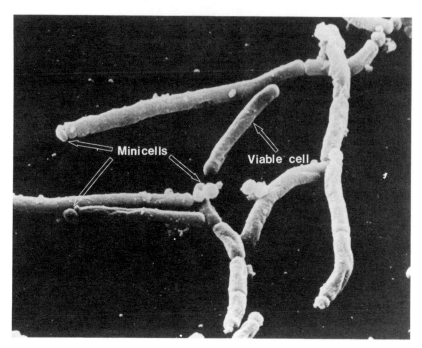

Figure 8.9
Electron micrograph showing the formation of minicells at the ends of viable cells. The cells shown here are *E. coli* K-12 X1776 containing pAU1, a recombinant plasmid carrying the rat insulin gene. (Ullrich *et al.*, Science 196:1313, 1977; courtesy of J. Shine and A. Ullrich.)

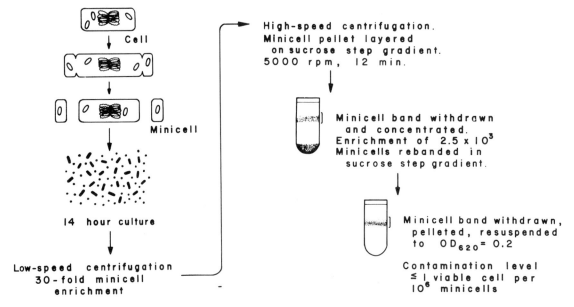

Figure 8.10
Diagrammatic representation of the preparation of *E. coli* minicells. Once the minicell preparations have been adjusted to the same optical density, they are incubated at 37°C in the presence of ^{35}S-methionine for 15 to 30 minutes. The cells are then washed free of the label, lysed, and analyzed by SDS-polyacrylamide gel electrophoresis.

thesized by minicells containing a recombinant plasmid and various derivatives of that plasmid can be used to establish the positions of genes on the recombinant plasmid.

The maxicell system involves the use of a special strain of *E. coli* that is sensitive to ultraviolet light. Irradiation of this strain results in such severe chromosomal damage that cell growth and chromosomally-directed protein synthesis cease. However, irradiation can be performed at a level that does not inactivate all of the plasmids in the cell. The irradiated maxicells remain capable of plasmid-directed polypeptide synthesis and will incorporate radioactive precursors into protein. As with minicells, these radioactive polypeptides can be examined by SDS-polyacrylamide gel electrophoresis to verify the expected protein products of various plasmid derivatives.

The coupled *in vitro* transcription-translation system is perhaps the most complicated of these three systems. The coupled system involves the preparation of a cell-free extract containing RNA polymerase, ribosomes, tRNA, and an energy-producing system. Under the proper reaction conditions, transcription and translation can be initiated by the addition of plasmid DNA. Because of the number of components involved and the sensitivity of these components to a variety of factors, the coupled systems require careful standardization of reaction conditions for optimal results. However, when used properly, this system can give detailed information concerning the regulatory aspects associated with the synthesis of gene products.

Systems that allow the determination of the protein products produced by various DNA fragments are valuable tools for the verification of the genetic results obtained by subcloning and other DNA manipulations.

References

Anderson, D., Shapiro, L., and Skalka, A. M. 1979. *In situ* immunoassays for translation products. In Methods in Enzymology, R. Wu, ed., 68:428–436.

Broome, S., and Gilbert, W. 1978. Immunological screening method to detect specific translation products. Proc. Natl. Acad. Sci. USA 75:2746–2749.

Bukhari, A. I., and Kamp, D. 1980. Genetic rearrangements and DNA cleavage maps. In Methods in Enzymology, L. Grossman and K. Moldave, eds., 65:436–449.

Chang, A. C. Y., Nunberg, J. H., Kaufman, R. J., Erlick, H. A., Schimke, R. T., and Cohen, S. N. 1978. Phenotypic expression in *E. coli* of a DNA sequence coding for mouse dihydrofolate reductase. Nature 275:617–624.

Clarke, L., and Clark, J. 1979. Selection of specific clones from colony banks by suppression or complementation tests. In Methods in Enzymology, R. Wu, ed., 68:396–408.

Clarke, L., Hitzeman, R., and Carbon, J. 1979. Selection of specific clones from colony banks by screening with radioactive antibody. In Methods in Enzymology, R. Wu, ed., 68:436–442.

Danna, K. J. 1980. Determination of fragment order through partial digests and multiple enzyme digests. In Methods in Enzymology, L. Grossman and K. Moldave, eds., 65:449–467.

CHAPTER 8
IDENTIFICATION AND CHARACTERIZATION OF RECOMBINANT TRANSFORMANTS

Dasgupta, S., and Inman, R. B. 1980. Denaturation mapping. In Methods in Enzymology, L. Grossman and K. Moldave, eds., 65:429–436.

Dunn, A. R., and Sambrook, J. 1980. Mapping viral mRNAs by sandwich hybridization. In Methods in Enzymology, L. Grossman and K. Moldave, eds., 65:468–478.

Erlich, H. A., Cohen, S. N., and McDevitt, H. O. 1979. Immunological detection and characterization of products translated from cloned DNA fragments. In Methods in Enzymology, R. Wu, ed., 68:443–453.

Grunstein, M., and Wallis, J. 1979. Colony hybridization. In Methods in Enzymology, R. Wu, ed., 68:379–389.

Hanahan, D., and Meselson, M. 1980. Plasmid screening at high colony density. Gene 10:63–67.

Heffron, F., Rubens, C., and Falkow, S. 1975. Translocation of a plasmid DNA sequence which mediates ampicillin resistance: Molecular nature and specificity of insertion. Proc. Natl. Acad. Sci. USA 72:3623–3636.

Holland, M. J., Holland, J. P., and Jackson, K. A. 1979. Cloning of yeast genes coding for glycolytic enzymes. In Methods in Enzymology, R. Wu, ed., 68:408–419.

Inselburg, J. 1977. Isolation, mapping, and examination of effects of TnA insertions in ColE1 plasmids. J. Bacteriol. 129:482–491.

Kaplan, D. A., Naumovski, L., and Collier, R. J. 1981. Chromogenic detection of antigen in bacteriophage plaques: a microplaque method applicable to large-scale screening. Gene 13:211–220.

Parker, R. C. 1980. Conversion of circular DNA to linear strands for mapping. In Methods in Enzymology, L. Grossman and K. Moldave, eds., 65:415–426.

Reiser, J., and Wardale, J. 1980. Sensitive immunological detection of translation products in SV40 plaques. Gene 12:11–16.

Roozen, K. J., Fenwick, R. G., and Curtiss, R. 1971. Synthesis of RNA and protein in plasmid containing minicells of *E. coli* K-12. J. Bacteriol. 107:21–23.

Ruvkun, G. B., Sundaresan, V., and Ausubel, F. M. 1982. Directed transposon Tn5 mutagenesis and complementation analysis of *Rhizobium meliloti* symbiotic nitrogen fixation genes. Cell 29:551–559.

Sancar, A., Hack, A., and Rupp, W. D. 1979. Simple method for identification of plasmid coded proteins. J. Bacteriol. 137:692–693.

Szostak, J. W., Stiles, J. I., Tye, B.-K., Chiu, P., Sherman, F., and Wu, R. 1979. Hybridization with synthetic oligonucleotides. In Methods in Enzymology, R. Wu, ed., 68:419–428.

Taub, F., and Thompson, E. B. 1982. An improved method for preparing large arrays of bacterial colonies containing plasmids for hybridization: *In situ* purification and stable binding of DNA on paper filters. Anal. Biochem. 126:222–230.

Woo, S. L. C. 1979. A sensitive and rapid method for recombinant phage screening. In Methods in Enzymology, R. Wu, ed., 68:389–395.

Zubay, G. 1973. *In vitro* synthesis of protein in microbial systems. Ann. Rev. Genet. 7:267–287.

EXERCISE 10

Phenotypic Characterization of His⁺ and Ara⁺ Recombinant Transformants

Materials

Luria Broth:
  2 agar (1.5%) plates containing 20 µg/ml ampicillin.
  2 agar (1.5%) plates containing 30 µg/ml tetracycline.
  2 agar (1.5%) plates containing 10 µg/ml tetracycline.

M9 Minimal Agar (Tc-M9 min.-ara): 1 plate supplemented with 0.2% arabinose, 166 µg/ml proline, 41 µg/ml leucine, 0.2 µg/ml thiamine hydrochloride (B_1), and 10 µg/ml tetracycline.

M9 Minimal Agar (Ap-M9 min.-hisol.): 1 plate supplemented with 0.2% glucose, 166 µg/ml proline, 0.2 µg/ml B_1, 20 µg/ml histidinol, and 20 µg/ml ampicillin.

Protocol

DAY 1

1. Assemble all Tc-M9 min.-ara and Ap-M9 min.-hisol plates used in Exercise 9, and record the approximate number of His⁺ and Ara⁺ transformants.

2. Using the scoring template provided on the back cover, patch as many transformants as possible on the selective media indicated below:

Transformation Plate →	Test Plate →	Master Plate
Ap-M9 min.-hisol (His⁺ transformants)	LB-Tc (30 µg/ml) (10 µg/ml)	Ap-M9 min.-hisol.
Tc-M9 min.-ara (Ara⁺ transformants)	LB-Ap (20 µg/ml)	Tc-M9 min.-ara

3. To score or patch the transformants, use a sterile loop, needle, or toothpick to transfer a few cells from a colony on the transformation plate to the test and master plates. Transfers should be made by gently touching the surface of the agar directly above the number on

EXERCISE 10

PHENOTYPIC CHARACTERIZATION OF HIS⁺ AND ARA⁺ RECOMBINANT TRANSFORMANTS

the template. Transfers can also be accomplished by making a short diagonal streak across the agar surface or by stabbing the cells into the agar. Avoid transferring too many cells since this can give rise to false colony formation.

Note: For the His⁺ transformants, two test plates are used. The first contains 30 µg/ml Tc while the second contains 10 µg/ml Tc. The desired His⁺ transformants should exhibit a Tc^r phenotype on 10 µg/ml Tc and a Tc^s phenotype on media containing 30 µg/ml Tc.

4. After completing the transfers, incubate the test and master plates at 37°C to allow for colony formation. Remember, minimal agar plates need to be incubated longer (2 to 4 days) than LB plates (1 to 2 days) to achieve good colony formation.

DAY 2

1. After all His⁺ and Ara⁺ transformants have been characterized with respect to their drug resistance phenotype, they can now be streaked for single colony isolates. Sector an LB-Ap plate (for His⁺ transformants) and an LB-Tc plate (for Ara⁺ transformants) into four equal quadrants (see fig. 2.4). As shown in Fig. 2.3, streak each quadrant with a different transformant. Incubate the plates at 37°C until single colonies are visible.

2. Make slants of the desired clones (Ap^r, His⁺, Tc^s at 30 µg/ml and Ap^s, Ara⁺, Tc^r) for future use and storage.

EXERCISE 11

Isolation of Recombinant Plasmid DNA and Retransformation of the Auxotrophic Host

Materials

SET Buffer: 20% sucrose, 50 mM Tris-HCl pH 7.6, 50 mM EDTA.

Lytic Mixture: 1% SDS, 0.2 N NaOH (make fresh mixture every two weeks).

Na(sodium) Acetate Buffer: 3.0 M, pH 4.8.

RNase Buffer: Pancreatic ribonuclease A, 10 mg/ml in 0.1 M Na acetate, 0.3 mM EDTA, pH 4.8, preheated to 80°C for 10 minutes.

Isopropanol.

Ethanol: 70%.

Water (ddH$_2$O): Sterile double distilled.

Luria Broth (LB): 8, 1 ml culture tubes.

Competent Cells: Use competent cells of BE42 and LA6 prepared in Exercise 9, Chapter 7.

LB Broth:
  4 agar (1.5%) plates containing 20 μg/ml ampicillin.
  4 agar (1.5%) plates containing 10 μg/ml tetracycline.

M9 Minimal Agar (Tc-M9 min.-ara): 1 plate supplemented with 0.2% arabinose, 166 μg/ml proline, 41 μg/ml leucine, 0.2 μg/ml thiamine hydrochloride (B$_1$), and 10 μg/ml tetracycline.

M9 Minimal Agar (Ap-M9 min.-hisol.): 1 plate supplemented with 0.2% glucose, 166 μg/ml proline, 0.2 μg/ml B$_1$, 20 μg/ml histidinol, and 20 μg/ml ampicillin.

Note: Using single colony isolates from the sectored plates prepared in Exercise 10, start 1 ml LB cultures of each of the desired transformants, at least 10 hours before the beginning of this exercise.

Protocol

DAY 1

1. Using the miniscreen technique for isolating plasmid DNA described in Exercise 5 (Chapter 3), purify plasmid DNA from each of the His$^+$ and Ara$^+$ transformants obtained.

EXERCISE 11

ISOLATION OF RECOMBINANT PLASMID DNA AND RETRANSFORMATION OF THE AUXOTROPHIC HOST

2. After the plasmid DNAs have been resuspended in 20 µl of ddH$_2$O, use 1 µl of each preparation to transform the appropriate auxotrophic host strain (i.e., BE42 and LA6). Perform the transformation according to Step 1, Day 2 of Exercise 9 (Chapter 7), substituting miniscreen DNA for pBR329 DNA. As a negative control, use 1 µg of pBR329 DNA to transform BE42 and LA6 competent cells.

3. After 3 hours at 37°C, spread 0.1 ml of each transformation mixture on either LB-Ap plates or LB-Tc plates (10 µg/ml). Incubate the plates at 37°C until colonies are visible.

DAY 2

1. Record the approximate number of Ap^r and Tc^r transformants obtained with each DNA preparation.

2. Patch at least 10 Ap^r transformants from each miniscreen DNA preparation onto an Ap-M9 min.-hisol. plate. Repeat the patching for Tc^r transformants onto Tc-M9 min.-ara plates. Incubate at 37°C.

3. Start restriction digests of miniscreen DNA preparations as described in Exercise 12.

EXERCISE 12

Physical Characterization of Recombinant Transformants: Restriction Endonuclease Mapping

Materials

(Restriction Endonuclease Digestion)

    *Pst*I Reaction Buffer (10X): 200 mM Tris-HCl pH 7.5, 100 mM $MgCl_2$, 500 mM $(NH_4)_2 SO_4$.

    *Hind*III Reaction Buffer (10X): 200 mM Tris-HCl pH 7.5, 70 mM $MgCl_2$, 500 mM NaCl, 70 mM 2-mercaptoethanol.

    *Bam*HI Reaction Buffer (10X): 200 mM Tris-HCl pH 7.0, 70 mM $MgCl_2$, 1 M NaCl, 20 mM 2-mercaptoethanol.

    *Bam*HI Endonuclease: Diluted to 1 unit/μl.

    *Pst*I Endonuclease: Diluted to 1 unit/μl.

    *Hind*III Endonuclease: Diluted to 1 unit/μl.

    Dilution Buffer: 10 mM Tris-HCl pH 7.5, 100 mM NaCl, 0.1 mM Na_2EDTA, 1 mM dithiothreitol (Cleland's reagent), 50% glycerol, 500 μg/ml BSA.

    Reaction Stop Mix: 5 M urea, 10% glycerol, 0.5% SDS, 0.025% xylene cyanole FF, 0.025% bromphenol blue WS.

    Miniscreen Plasmid DNA: Isolated in Exercise 11.

    1 M Tris-HCl pH 7.5.

    1 M NaCl.

    Bacteriophage Lambda DNA: Molecular size standards prepared in Exercise 6.

    Water (ddH_2O): Sterile double distilled.

    Ice.

(Agarose Gel Electrophoresis)

    Tris-Borate Buffer (1X): 90 mM Tris-HCl pH 8.2, 2.5 mM Na_2 EDTA, 89 mM boric acid.

    Agarose: Molten, 1% solution made in Tris-Borate Buffer (1X).

    Ethidium Bromide: 4 μg/ml, made with water.

Note: To construct restriction enzyme cleavage maps of the His$^+$ and Ara$^+$ plasmids, the following assumptions should be made.

EXERCISE 12
PHYSICAL CHARACTERIZATION OF RECOMBINANT TRANSFORMANTS: RESTRICTION ENDONUCLEASE MAPPING

1. A plasmid that confers a His$^+$ phenotype to BE42 carries a fragment of DNA inserted in the *Hin*dIII site of the vector. Plasmids that confer an Ara$^+$ phenotype to LA6 carry a *Pst*I fragment in the corresponding site of pBR329.

2. The *Hin*dIII insert possesses one *Pst*I site located 34% from the left-hand end of the fragment. The *Pst*I insert possesses two *Bam*HI sites located approximately 35% and 91% from the left end of the DNA fragment.

3. Neither insert contains an *Eco*RI cleavage site.

Protocol

DAY 1

1. Using one of the miniscreen DNA preparations made from His$^+$ transformants in Exercise 11, set up the following restriction digests:

Reaction 8-1a	μl	Reaction 8-1b	μl
His$^+$ miniscreen DNA	1.5	His$^+$ miniscreen DNA	1.5
1 unit *Hin*dIII endo.	1.0	1 unit *Pst*I endo.	1.0
10X *Hin*dIII Buffer	1.5	10X *Pst*I Buffer	1.5
ddH$_2$O	11	ddH$_2$O	11
	15		15

2. Using one of the miniscreen DNA preparations made from Ara$^+$ transformants in Exercise 11, set up the following restriction digests:

Reaction 8-2a	μl	Reaction 8-2b	μl
Ara$^+$ miniscreen DNA	1.5	Ara$^+$ miniscreen DNA	1.5
1 unit *Pst*I endo.	1.0	1 unit *Bam*HI endo.	1.0
10X *Pst*I Buffer	1.5	10X *Bam*HI Buffer	1.5
ddH$_2$O	11	ddH$_2$O	11
	15		15

3. Transfer all reaction tubes to 37°C and incubate for 1 hour. At the end of 1 hour, take 7.5 μl from each reaction and add it to a microfuge tube containing 3 μl of reaction stop mix. Store the remaining portion of each reaction at −20°C.

4. Set up a 1% agarose submarine minigel or vertical slab gel as shown in Figs. 5.4 and 5.5, Chapter 5. Once the gel is submerged in 1X Tris-Borate Buffer, load the samples just above the surface of the sample-loading wells. Load the reactions and volumes indicated at the top of the next page.

—Gel 1—

Well No.:	1	2	3	4	5	6	7	8	9	10	11	12
Reaction:		8–1a	8–1b		8–2a	8–2b						
Molecular Weight Markers:				4–10								
µl loaded: (Vert. slab)		10	10	10	10	10						
µl loaded: (minigel)		10	10	5	10	10						

Note: The *Hin*dIII cleaved λ DNA prepared in Exercise 6 will serve as molecular weight markers.

5. If the DNA samples have been cleaved, set up the following restriction digests:

Reaction 8–4a	µl	*Reaction 8–4b**	µl
*Pst*I cleaved Ara$^+$ miniscreen DNA	7	*Bam*HI cleaved Ara$^+$ miniscreen DNA	7
1 unit *Bam*HI endo.	1.5	1 unit *Eco*RI endo.	1
1 M NaCl	1.5	1 M Tris-HCl, pH 7.5	1
ddH$_2$O	5.5	ddH$_2$O	6
	15		15

*An *Eco*RI digestion of *Bam*HI cleaved Ara$^+$ miniscreen DNA will help resolve two comigrating *Bam*HI fragments.

6. Incubate all reaction mixtures at 37°C for approximately 1 hour. Take 7 µl from each reaction and add it to a microcentrifuge tube containing 3 µl of reaction stop mix. Store the remaining portions of the reactions at −20°C.

7. Run the terminated reactions with molecular weight markers on a 1% agarose gel (Gel 2) as described in Step 4, Day 1, of this exercise.

8. Using a ruler and the photographic prints or negatives of Gels 1 and 2, determine the approximate size of the restriction fragments. Use this information to establish the orientation of the DNA inserts (relative to restriction sites on the vector) as well as the approximate positions of restriction sites on the inserts themselves. The restriction map and nucleotide sequence of pBR329 (Appendix 3) can be used to help position restriction sites of the cloned inserts.

EXERCISE **13**

Determination of Histidinol Dehydrogenase Activity in His⁺ Transformants

Materials

Dye Mix: 5 parts 2-*p*-iodophenyl-3-*p*-nitrophenyl-5-phenyltetrazolium chloride (INT), 1 part phenazine methosulfate (PMS), 1 part gelatin, and 1 part 20 mg/ml NAD, pH ~ 5.5 (stored at −20°C).

0.1 M Histidinol-HCl; pH 8.0.

0.1 M Triethanolamine-HCl; pH 7.5.

1 M Diethanolamine.

0.67 N HCl.

Toluene.

Ice.

Note: Dye mix preparation: Make fresh each day in brown bottle and keep at 4°C in the dark.

1. Dissolve 3.2 mg/ml INT in water warmed almost to boiling (store refrigerated in brown bottle).
2. Dissolve (must be warmed) 0.4 mg/ml PMS and store at 4°C in brown bottle. (Both solutions will keep several months.)
3. Boil briefly to dissolve 0.2% gelatin (store refrigerated).
4. Mix reagents in proportions described above.

Histidinol dehydrogenase is the enzyme involved in catalyzing the last step in histidine biosynthesis. It is coded for by the *his*D gene and is presumed to catalyze the conversion of histidinol to enzyme-bound histidinol and then the conversion of the latter to histidine (histidinol has never been isolated as an intermediate, however). This reaction requires the reduction of two moles of NAD per mole of histidinol.

The assay for histidinol dehydrogenase is a colorimetric assay in which the reduced NAD formed in the enzymatic reaction is used to reduce the coupled dye system, PMS with INT, to form a red INT-formazan. In this assay, a unit of enzyme yields 1.0 absorbance unit at 520 mμ. Specific activity is in terms of units of enzyme per absorbance unit of cell suspension at 650 nm.

Protocol

DAY 1

(a) Toluenization:

1. Take 1.5 ml samples from each of the following M-9 minimal glucose cultures: *E. coli* K12-AB257 (His$^+$); *E. coli* K12, BE42 (His$^-$); BE42 (His$^+$). Place samples in 15 ml Corex tubes and spin cells down at 10 krpm, 5 minutes at 4°C.

2. Resuspend cell pellets in 3 ml of 0.1 M triethanolamine-HCl buffer, pH 7.5. Spin cells down as before and resuspend pellets in 1.5 ml of the same buffer.

3. Take 0.5 ml of the cell suspension and make a spectrophotometer reading at 650 nm. Add 0.01 ml toluene to the remaining cell suspension and shake tubes in rotary shaker for 20 minutes at 37°C. Cool tubes in ice bath.

(b) Enzyme assay:

1. Add 0.07 ml of 1 M diethanolamine to the 1.0 ml of toluenized cells to bring the pH to 8.6.

2. Put 0.29 ml of the cell suspension into each of 2 tubes and add 0.2 ml of "dye mix" to each tube.

3. Add 10 µl histidinol to one of the tubes and 10 µl water to the other; label the second tube "blank." Incubate at 37°C for 20 minutes.

4. Add 0.1 ml of 0.67 N HCl to stop the reaction. Read assay mixture against the blank at 520 nm with the spectrophotometer. Determine the enzymatic activity by calculating the A_{520}/A_{650} ratio.

After calculating the specific activity of the histidinol dehydrogenase activity present in each of the samples, compare the levels present in the wild type AB257 (His$^+$) and the mutant BE42 (His$^-$) with the level present in cells containing your recombinant plasmids, BE42 (His$^+$). What conclusions can be drawn concerning the effects of the recombinant plasmid on the ability of the cell to express histidinol dehydrogenase activity?

EXERCISE 13

DETERMINATION OF HISTIDINOL
DEHYDROGENASE ACTIVITY
IN HIS$^+$ TRANSFORMANTS

CHAPTER 9

Cloning Regulatory DNA Sequences

As described in the introductory section of the last chapter, gene localization (or mapping) is one of the primary objectives of molecular cloning. Indeed, if a gene of interest is to be fully understood or exploited for practical purposes, a fine structure map of the gene must be constructed and the regulatory features of the gene determined. While applying the techniques of subcloning and maxicell analysis to a recombinant plasmid may yield the approximate size and location of a gene, they do not usually reveal the direction of transcription or information about the factors controlling gene expression. This information must be obtained by analyzing the regulatory sequences for that gene.

Regulatory DNA Sequences

Regulatory DNA sequences are noncoding regions of the DNA that determine if, when, and at what level a particular gene is expressed. The genetic information contained in the regulatory sequence manifests itself through specific interactions with regulatory or other cellular proteins. The following are of some examples of common prokaryotic regulatory DNA sequences:

1. Promoter (point of transcription initiation)
2. Terminator (point of transcription termination)
3. Operator (point of repressor protein binding)
4. Catabolite activator-protein (CAP) binding site

In eukaryotic organisms, other types of regulatory sequences have been found. For example, within those genes transcribed by RNA polymerase III, transcription factor binding sites have been identified that are required for accurate and efficient transcription. Because these sites can act to initiate transcription without what would normally be considered

CHAPTER 9
CLONING REGULATORY DNA SEQUENCES

the promoter region (i.e., 5' flanking sequences), they have been loosely termed "internal promoters." Another interesting eukaryotic regulatory DNA sequence is the "enhancer." These sequences can act at a distance, sometimes thousands of base pairs away, to enhance transcription from an existing promoter.

Until recently, regulatory DNA sequences had to be examined *in vitro*, using sophisticated biochemical techniques. Genetic approaches to the study of gene regulation frequently provided indirect and sometimes ambiguous results. However, with the development of recombinant DNA technology and the construction of specialized cloning vectors, regulatory sequences can now be isolated and examined *in vivo*. The following discussion will be primarily concerned with prokaryotic promoters and promoter cloning plasmid vectors.

Promoter Structure and Function

In *E. coli*, transcription can be divided into three distinct stages: initiation, elongation, and termination. The enzyme involved in each of these stages is RNA polymerase. The enzyme is an aggregate protein consisting of a core and an additional factor called sigma. It is the sigma subunit that confers transcriptional specificity to the complex. In the absence of the sigma subunit, the core enzyme binds to and transcribes double-stranded DNA indiscriminantly.

Promoter selection is the first step in the initiation process. Current models for promoter recognition suggest that the holoenzyme diffuses to a promoter site, where some signal in the DNA base sequence allows the polymerase to form a specific but transient "closed complex." Next, a localized denaturation of DNA helix occurs and the complex undergoes a change to form an "open complex." Finally, when the open complex is supplied with ATP or GTP (sometimes CTP), RNA synthesis initiates.

The nucleotide sequences of more than 60 prokaryotic promoter regions have been determined to date (see Rosenberg and Court, 1979, for a review). On the basis of sequence analysis, a promoter can be defined as a particular arrangement of 40 to 50 base pairs of DNA, which is recognized, bound, and transcribed by RNA polymerase. Within this sequence of DNA, three structural domains have been identified by genetic and biochemical studies: (1) position 1, the purine initiation nucleotide from which RNA synthesis begins; (2) position -6 to -12, the Pribnow box or firm binding site for RNA polymerase; (3) position -35, a region thought to be involved in the initial recognition of the promoter. Using computer-aided analysis of several promoter sequences, these structural similarities or domains can be represented as a single "consensus sequence" (fig. 9.1a). This sequence also illustrates two important points: (1) while all promoters are different, they share certain structural and functional features; and (2) promoters are relatively large (>40 bp), highly organized regions of DNA and the probability that such a structure would appear by chance in the genome is very low.

It should be stressed that the consensus sequence shown in Fig. 9.1a represents an average of several different promoters. The strength of these promoters ranges from weakly constitutive to strongly inducible with the levels of transcription being controlled by a variety of

a. Prokaryotic RNA Pol (E. coli)

TTGACA ← 16-19 bp → TATAATG ← 4-7 bp → CAT*

−35 region   Pribnow box

b. Eukaryotic RNA Pol II

GG^C_TCAATCT ← 35-40 bp → GTATA$^{AA}_{TAT}$G··G ← 9-17 bp → Py···PyAPy*

TATA box

Figure 9.1
(a) Consensus sequence for bacterial promoters. The sequence and spacing shown are those found most frequently in the promoter sequences analyzed (Rosenberg and Court, 1979). (b) Consensus sequence for eukaryotic promoters (Breathnach and Chambon, 1981).

CHAPTER 9
CLONING REGULATORY DNA SEQUENCES

mechanisms (e.g., catabolite repression and attenuation). Therefore, it is not surprising that a synthetic oligonucleotide of the consensus sequence behaved as a promoter of average strength when cloned and examined in *E. coli*.

Although a conceptual model of eukaryotic promoter structure is developing, the details are currently less well-established than is the case with prokaryotic promoters. Transcription from eukaryotic RNA polymerase II promoters appears to initiate at an A residue that is surrounded by pyrimidines in the 5′ flanking region of the gene (fig. 9.1b). An AT-rich region of homology located 25 to 30 base pairs upstream from the transcription initiation point has been termed the Goldberg-Hogness box or "TATA" box. This region is similar to the Pribnow box of prokaryotic promoters. In addition, a 9 base pair conserved sequence has been observed 70 to 80 base pairs from the transcription initiation points of several cellular and viral protein genes. However, the significance of this sequence to the general structure of eukaryotic promoters has not been established.

When prokaryotic RNA polymerase is bound to a promoter, a region of approximately 55 base pairs of DNA is protected from interaction with other proteins. When a restriction endonuclease cleavage site occurs within this protected region of DNA, as long as the RNA polymerase molecule remains tightly bound to the DNA, the restriction enzyme cannot cleave at its recognition site. If a plasmid molecule can be cleaved at five sites in the absence of RNA polymerase but at only four sites in the presence of RNA polymerase, it can be concluded that the binding of polymerase protects the fifth site, suggesting that this site occurs in the vicinity of an RNA polymerase binding site. The location of the protected site can be determined by examining the digestion products in the presence and absence of RNA polymerase. For example, assume that there are five *Hin*dIII cleavage sites in a plasmid and that digestion gener-

RNA Polymerase Binding Studies

ates fragments of 0.5, 1.0, 2.0, 2.5, and 3.0 megadaltons, while digestion in the presence of RNA polymerase generates fragments of 1.0, 2.0, 2.5, and 3.5 megadaltons. The binding of polymerase prevents the cleavage of the HindIII site between the 0.5 and 3.0 megadalton DNA fragments, suggesting the existence of a promoter within 50 base pairs of this cleavage site. As shown in Fig. 9.2, RNA polymerase protection experiments can be used to show that the single HindIII site in pBR322 is located in or near a plasmid promoter. Subsequent results have demonstrated that this site actually occurs within the promoter for the tetracycline resistance gene.

Protection experiments are useful only when a restriction cleavage site occurs within a promoter region. Filter binding assays, on the other hand, can be used to detect the formation of RNA polymerase-DNA complexes without the requirement for a convenient restriction site. Double-stranded DNA will not bind efficiently to a nitrocellulose filter, but a DNA-protein complex will bind to nitrocellulose under appropriate conditions. Most promoter-containing DNA fragments will form filter-

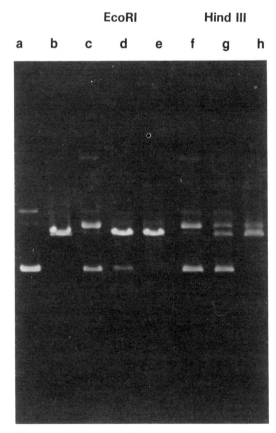

Figure 9.2
RNA polymerase protection of the HindIII site of pBR322. Supercoiled pBR322 was incubated with RNA polymerase, then digested with EcoRI or HindIII. Lane a contains undigested plasmid, and lane b contains linear plasmid. RNA polymerase has been bound to the DNA in lanes c and f. Lanes d and g illustrate how the binding of polymerase prevents cleavage by the restriction enzymes. In lanes e and h, this protection has been eliminated by the addition of all four ribotriphosphates to allow the polymerase to move during transcription.

CHAPTER 9

CLONING REGULATORY
DNA SEQUENCES

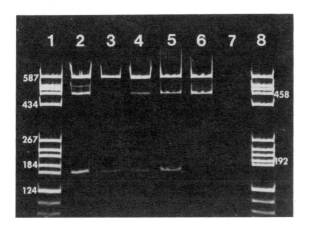

Figure 9.3
An RNA polymerase/nitrocellulose filter binding study of the promoter-containing *Hae*III restriction fragments of pPV2, a plasmid derivative of pBR322. The digested plasmid DNA was incubated with RNA polymerase and passed over nitrocellulose filters. The filter-retained DNA was eluted and analyzed by electrophoresis on a 7.5% polyacrylamide gel. The *Hae*III fragments of pPV2 and pBR322 are shown in Lanes 1 and 8, respectively. Lanes 2 to 7 show the promoter-containing restriction fragments retained on nitrocellulose filters under different conditions of RNA polymerase binding. For example, the reaction condition for Lane 2 contained low salt (KCl) and excess RNA polymerase, while the condition used in Lane 3 contained high salt and low RNA polymerase. The numbers adjacent to Lanes 1 and 8 are the sizes, in bp, of some of the *Hae*III restriction fragments.

retainable complexes with RNA polymerase. In order to utilize this technique in the localization of promoters, a mixture of DNA fragments is allowed to interact with RNA polymerase, then the mixture is passed through a nitrocellulose filter. The filtrate can be collected, then the bound material eluted from the filter and collected. The two samples can be examined by gel electrophoresis to determine whether any of the DNA fragments have been retained by the filter as DNA-protein complex (fig. 9.3). Although the factors contributing to the formation of filter-retainable complexes have not been completely determined, the degree of filter-retention is generally proportional to the strength of RNA polymerase binding.

Terminators

Transcriptional terminators play a vital role in the regulation of gene expression by establishing boundaries for specific transcriptional units within the genome. Without those boundaries, readthrough transcription into adjacent regions would occur. This would place all genes on the chromosome under the control of a few strong promoters. In order for a gene or operon to maintain its particular regulatory properties, readthrough transcription from adjacent genes must be prevented by the action of terminator sequences.

A number of terminator sequences from various prokaryotic organisms have been analyzed. Like promoters, three structural domains have been identified: (1) a region of hyphenated dyad symmetry (i.e., an inverted repeat sequence) preceding the termination site; (2) uracil (U) resides at the 3' end of the RNA transcript; and (3) a G/C rich sequence of variable length surrounding the hyphenation of the dyad symmetry and preceding the termination site. The latter feature makes possible the formation of a "stem-loop" structure at the end of the DNA molecule. The G/C richness in the pre-termination region also provides the potential to form stable RNA:DNA as well as RNA:RNA base pairs. However, it must be stressed that although these sequences are found in those regions of DNA where RNA transcripts terminate, it is still unclear if and how these sequences are involved in the termination event.

At least three mechanisms of transcriptional termination have been observed in prokaryotes:

1. Rho-independent termination in which transcription is terminated at a defined spot. (These terminators generally contain the GC-rich regions of hyphenated dyad symmetry as described above.)

2. Rho-dependent termination that is functional only when translation of mRNA is inhibited. Rho-dependent terminators are thought to exist within operons and to be responsible for polarity when nonsense mutations are introduced into the operon. These terminators contain dyad symmetries that are AT-rich, are not followed by T residues, and termination occurs heterogeneously in an AT-rich region.

3. Attentuation of gene expression, as is the case with the *trp* operon. The rate of translation of a leader RNA sequence rich in tryptophan codons affects the ability of RNA polymerase to continue transcription into the *trp* operon. Because the rate of translation of the leader sequence is affected by the level of charged $tRNA_{trp}$ in the cell, termination of transcription at the attenuator sequence provides a system for modulating the expression of the genes involved in the synthesis of tryptophan.

The eukaryotic terminators that have been examined appear to be similar to the rho-dependent terminators of prokaryotes. Termination occurs heterogeneously in a region of AT-rich dyad symmetry. Based on the sequence analysis of the 3' untranslated region of fifteen genes, it has been suggested that the structure

$$(\text{T-rich}) \cdots \text{TAG} \cdots \genfrac{}{}{0pt}{}{\text{TAGT}}{\text{TATGT}} \cdots (\text{AT-rich}) \cdots \text{TTT}$$

may play an important role in transcription termination in yeast.

Promoter and Terminator-Probe Plasmid Cloning Vectors

As mentioned in Chapter 1, promoter-probe and terminator-probe plasmid vectors have been constructed to facilitate the isolation and characterization of regulatory DNA sequences *in vivo*. Although each of these vectors is unique in terms of method of construction and genetic markers, they all utilize the same principle of "insertional activation" of a transcriptionally silent plasmid-encoded gene. The plasmid gene to be activated is usually made transcriptionally silent by deleting its promoter, leaving in its place a unique restriction enzyme cleavage site. Activation of the silent gene occurs when a promoter-active restriction fragment is inserted in the correct orientation into the unique cleavage site. The inserted promoter initiates transcription that proceeds through the plasmid gene (readthrough transcription). The resulting readthrough transcript contains a functional translational initiation site and coding sequence of the silent gene, and can be translated to produce a functional gene product. Upon transformation of the appropriate bacterial host, the phenotype of the promoter-active recombinant plasmid can be selected for directly on the appropriate medium.

An example of a promoter-probe plasmid vector is shown in Fig. 9.4. This plasmid, pPV33, is a derivative of pBR327, which carries a 32 base pair deletion in the promoter of the tetracycline resistance (Tc^r) gene. Cells containing pPV33 exhibit a tetracycline sensitive, ampicillin resistant (Tc^s, Ap^r) phenotype. When a population of randomly generaged *Eco*RI fragments is cloned into the *Eco*RI site of pPV33, only promoter-active fragments are capable of activating the silent Tc^r gene. Promoter-active recombinants can be detected by screening Ap^r transformants for growth on media containing a range of Tc concentrations. The relative level of Tc^r can be taken as an approximation of promoter strength. The level of antibiotic resistance is often expressed as the efficiency of plating (EOP_{50}) of a resistant strain. The EOP_{50} value is a common clinical measure of antibiotic resistance that simply refers to the concentration of antibiotic that inhibits cell growth by 50 percent. By plating a dilution of drug-resistant bacterial culture on nutrient agar and nutrient agar containing increasing amounts of antibiotic, a concentration should be found that gives half the number of colonies observed on the drug-free medium. This concentration of antibiotic represents the EOP_{50} value for that strain.

An example of a promoter-active recombinant of pPV33 is shown in Fig. 9.5. The plasmid pLP1 contains the wild-type promoter/operator region from the *lac* operon of *E. coli* as well as remnants of the I (repressor) and Z (β-galactosidase) genes. In the configuration shown, the Tc^r gene is expressed in *E. coli*. Moreover, the level of Tc^r is higher on media containing arabinose as a carbon source than on glucose-containing media, as would be expected if the Tc^r gene in pLP1 is controlled by catabolite-repressed transcription from the *lac* promoter.

CHAPTER 9

CLONING REGULATORY DNA SEQUENCES

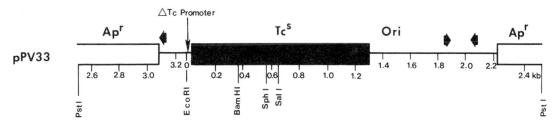

Figure 9.4
Genetic and restriction map of the promoter-probe plasmid vector pPV33. The circular plasmid map has been opened at the unique *Pst*I cleavage site. Numbers below the linear map are kilobase (kb) coordinates. Arrows represent the approximate location and direction of transcription of plasmid promoters. ΔTc promoter indicates the location of a 32 base pair deletion in the promoter for the tetracycline resistance gene.

CHAPTER 9

CLONING REGULATORY
DNA SEQUENCES

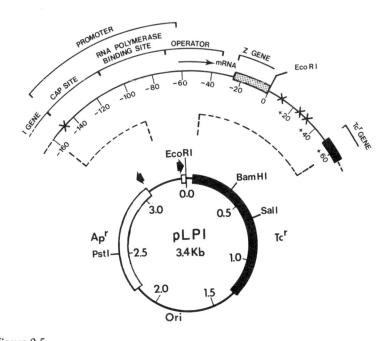

Figure 9.5
The circular restriction and genetic map of pLP1. The map is drawn in approximate scale with the Ap^r gene (open bar), Tc^r gene (solid bar), and truncated *lacZ* gene (stippled bar) indicated. Thick arrows indicate the location and orientation of the Ap^r and wild-type *lac* promoters. The region surrounding the unique *Eco*RI site (position 0.0) has been expanded. To the left of the *Eco*RI site lies the *lac* promoter, operator, and remnants of the *lacI* gene and *lacZ* gene. The point of initiation and direction of transcription from the *lac* promoter are indicated with a thin arrow. To the right of the *Eco*RI site are located 3 "out-of-phase" nonsense codons (indicated by "X"). Map intervals of the pLP1 circular map are provided in kb, while those of the expanded region are given in bp.

Because the product of the Tc^r gene is a membrane-bound protein as opposed to an antibiotic-inactivating enzyme, measurement of the Tc^r levels in promoter-active recombinants is more of a qualitative than quantitative estimate of promoter strength. For this reason, additional promoter-cloning vectors have been constructed that utilize silent marker genes that encode enzymes for which reliable quantitative assays are available. For example, genes for the enzymes β-galactosidase, galactokinase, and chloramphenicol acetyltransferase (CAT) have been incorporated into the promoter-probe vectors pMC81, pKO-1, and pPV501, respectively. In the case of the latter vector, the chloramphenicol resistance (Cm^r) gene was obtained from pBR328 as a 778 base pair *Taq*I fragment and inserted into the *Sal*I site of pPV33 (fig. 9.6). With the plasmid pPV501, the Cm^r gene can be activated by promoter-active fragments inserted at the *Sal*I, *Sph*I, *Bam*HI and *Eco*RI cleavage sites. Unlike pPV33, the strength of promoter-active fragments cloned in pPV501 can be measured in terms of units of CAT activity instead of EOP$_{50}$.

Terminator-probe plasmid vectors use the properties of readthrough transcription in the inverse manner. In these plasmids, a unique cloning site has been introduced between a promoter and the translational start

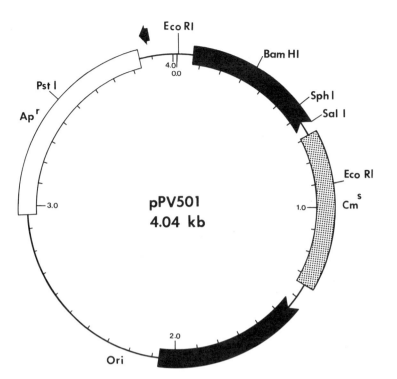

Figure 9.6
Circular genetic and restriction map of the promoter-probe vector pPV501. Kilobase (kb) coordinates are provided inside the circular map. The Ap^r gene, inactivated Tc^r gene, and silent Cm^r gene are indicated as open, solid, and stippled bars, respectively. The promoter for the Ap^r gene is indicated by the thick arrow.

CHAPTER 9

CLONING REGULATORY
DNA SEQUENCES

point of a marker gene. The insertion of a DNA fragment that contains a transcriptional terminator can block readthrough transcription into the marker gene, thus decreasing or eliminating its expression. An example of a simple terminator-probe vector is shown in Fig. 9.7. The plasmid pCP1 (a promoter-active recombinant plasmid derived from pPV33) contains the promoter for the Cm^r gene of pBR328 inserted next to the silent Tc^r gene of pPV33. As is the case with pLP1, the Tc^r gene is expressed by readthrough transcription from the CAT promoter. As indicated in the figure, terminator-containing DNA fragments can be inserted into the unique *Eco*RI site to terminate transcription into the Tc^r gene. Therefore, terminator-active recombinant plasmids of pCP1 can be detected *in vivo* by scoring Ap^r transformants for reduced levels of Tc^r. Unfortunately, Tc^s recombinants of pCP1 and other terminator-probe plasmids can be isolated that do not contain legitimate terminators cloned in the *Eco*RI site. This can happen when a cloned DNA fragment simply confers some unusual structure to the hybrid mRNA that reduces the stability or the efficiency of translation of the readthrough transcript. Alternatively, the cloned fragment may merely contain translational stop codons (TAA, TAG, and TGA) that serve to uncouple translation (from the insert) from readthrough transcription by a mechanism similar to rho-dependent termination. In some cases, uncoupling of the two pro-

CHAPTER 9

CLONING REGULATORY DNA SEQUENCES

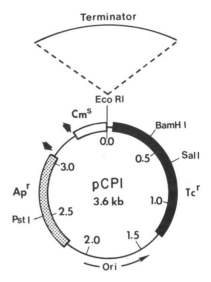

Figure 9.7
Genetic and restriction map of the terminator-probe plasmid vector pCP1. Kilobase (kb) coordinates are given on the inside of the circular map. The Ap^r gene, Tc^r gene, and beginning portion of the Cm^r gene are indicated by stippled, solid, and open bars, respectively. The thick arrows indicate the promoters for the Ap^r and Cm^r genes. Ori represents the origin and direction of plasmid DNA replication.

cesses also results in the termination of transcription. In addition, the production of fusion polypeptides as a result of translation initiation events originating within the cloned insert may interfere with the production of the active marker gene product. However, the vectors pCP1 and pKO-1 (which can also be used as a terminator-probe vector), have three out-of-phase translational stop codons immediately preceding the beginning of the marker gene (fig. 9.5) to prevent the production of fusion proteins. While terminator-cloning vectors can be used to clone real terminator sequences, the user should be aware of potential "pseudoterminator" effects that can yield artifactual results.

The construction of specialized cloning vectors is a significant development in the study of the control of gene expression. The use of these vectors allows for the isolation and characterization of regulatory DNA sequences on small fragments of DNA. Regulatory sequences that are difficult to study separately can be linked to marker genes that can be easily assayed both *in vivo* and *in vitro*. The use of the specialized vectors can provide the biological evidence needed to demonstrate the involvement of a DNA sequence in the transcriptional process.

References

Adhya, S., and Gottesman, M. 1978. Control of transcription termination. Ann. Rev. Biochem. 47:967–996.

Breathnach, R., and Chambon, P. 1981. Organization and expression of eucaryotic split genes coding for proteins. Ann. Rev. Biochem. 50:349–383.

Calva, E., and Burgess, R. R. 1980. Characterization of a rho-dependent termination site within the *cro* gene of bacteriophage λ. J. Biol. Chem. 255:11017–11022.

Close, T. J., and Rodriguez, R. L. 1982. Construction and characterization of the chloramphenicol-resistance gene cartridge: A new approach to the transcriptional mapping of extrachromosomal elements. Gene 20:305–316.

Frunzio, R., Bruni, C. B., and Blasi, F. 1981. *In vivo* and *in vitro* detection of the leader RNA of the histidine operon of *Escherichia coli* K-12. Proc. Natl. Acad. Sci. USA 78:2767–2771.

Greenfield, L., Boone, T., and Wilcox, G. 1978. DNA sequence of the *ara* BAD promoter in *Escherichia coli* B/r. Proc. Natl. Acad. Sci. USA 75:4724–4728.

Kolter, R., and Yanofsky, C. 1982. Attenuation in amino acid biosynthesis operons. Ann. Rev. Genet. 16:113–134.

Küpper, H., Sekiya, T., Rosenberg, M., Egan, J., and Landy, A. 1978. A rho-dependent termination site in the gene coding for tyrosine tRNA su_3 of *Escherichia coli.* Nature 272:423–428.

McKenney, K., Shimatake, H., Court, D., Schmeissner, U., Brady, C., and Rosenberg, M. A system to study promoters and terminator signals recognized by *Escherichia coli* RNA polymerase. In Gene Amplification and Analysis, Vol. II: Structural Analysis of Nucleic Acids, J. G. Chirikin and T. S. Papas, eds., Elsevier/North-Holland, New York, 1981.

Rosenberg, M., and Court, D. 1979. Regulatory sequences involved in the promotion and termination of RNA transcription. Ann. Rev. Genet. 13:319–353.

Rosenberg, M., McKenney, K., and Schümperli, D. Use of the *E. coli* galactokinase gene to study prokaryotic and eukaryotic gene control signals. In Promoters: Structure and Function, R. L. Rodriguez and M. J. Chamberlin, eds., Praeger Scientific, New York, 1982.

Seeburg, P. H., and Schaller, H. 1975. Mapping and characterization of promoters in bacteriophage fd, f1 and M13. J. Mol. Biol. 92:261–277.

Shaw, W. V. 1975. Chloramphenicol acetyltransferase from chloramphenicol-resistant bacteria. Meth. Enzymol. 43:737–755.

West, R. W., and Rodriguez, R. L. 1980. Construction and characterization of *E. coli* promoter-probe plasmid vectors II. RNA polymerase binding studies on antibiotic resistance promoters. Gene 9:175–193.

West, R. W., Kline, E. L., Horwitz, A. H., Wilcox, G., and Rodriguez, R. L. Cloning and analysis of inducible *Escherichia coli* promoters using promoter-probe plasmid vectors. In Promoters: Structure and Function, R. L. Rodriguez and M. J. Chamberlin, eds., Praeger Scientific, New York, 1982.

West, R. W., and Rodriguez, R. L. 1982. Construction and characterization of *E. coli* promoter-probe plasmid vectors III. pBR322 derivatives with deletions in the tetracycline resistance promoter region. Gene 20:291–304.

Zaret, K. S., and Sherman, F. 1982. DNA sequence required for efficient transcription termination in yeast. Cell 28:563–573.

CHAPTER 9

CLONING REGULATORY DNA SEQUENCES

EXERCISE 14

Cloning the Promoters for the Arabinose and Histidine Operons of E. coli

Materials

SET Buffer: 20% sucrose, 50 mM Tris-HCl pH 7.6, 50 mM EDTA.

Lytic Mixture: 1% SDS, 0.2 N NaOH (make fresh every two weeks).

K (potassium) Acetate Buffer: 3.0 M, pH 4.8.

Phenol: Buffer saturated.

Isopropanol.

Ethanol: 70%.

Glycerol: 80%.

Chloroform.

2 M NaCl.

Ice.

Luria Broth (LB): 50 ml in a sterile 100 ml bottle.

0.3 M Sterile $CaCl_2$.

Competent RRI Cells (see Exercise 9).

Luria Broth:
  5 agar (1.5%) plates containing 20 μg/ml ampicillin.
  4 agar (1.5%) plates containing 10 μg/ml tetracycline.
  4 agar (1.5%) plates containing 10 μg/ml chloramphenicol.

*Eco*RI Endonuclease: Diluted to 1 unit/μl.

*Bgl*II Endonuclease: Diluted to 1 unit/μl.

*Bam*HI Endonuclease: Diluted to 1 unit/μl.

T4 DNA ligase: Diluted to 1 unit/μl.

*Eco*RI Reaction Buffer (10X): 1 M Tris-HCl pH 7.5, 50 mM $MgCl_2$, 20 mM 2-mercaptoethanol, 1 M NaCl.

*Bgl*II Reaction Buffer (10X): 1M Tris-HCl pH 8.0, 50 mM $MgCl_2$, 1 M NaCl.

*Bam*HI Reaction Buffer (10X): 200 mM Tris-HCl pH 7.0, 1 M NaCl, 70 mM $MgCl_2$, 20 mM 2-mercaptoethanol.

EcoRI* Reaction Buffer (10X): 200 mM Tris-HCl pH 8.8, 20 mM MgCl$_2$, 50 mM NaCl, 20 mM 2-mercaptoethanol.

Ligase Reaction Buffer (10X): 200 mM Tris-HCl pH 7.5, 100 mM MgCl$_2$, 100 mM dithiothreitol.

ATP (10X): 5 mM.

Reaction Stop Mix: 5 M urea, 10% glycerol, 0.5% SDS, 0.025% xylene cyanole FF, 0.025% bromophenol blue WS.

EXERCISE 14

CLONING THE PROMOTERS FOR THE ARABINOSE AND HISTIDINE OPERONS OF *E. COLI*

Agarose Gel Electrophoresis

Tris-Borate Buffer (1X): 90 mM Tris-HCl pH 8.2, 2.5 mM Na$_2$EDTA, 89 mM boric acid.

Agarose: Molten, 1% made in Tris-Borate Buffer (1X).

Ethidium Bromide: 4 µg/ml made with H$_2$O.

Purified Plasmid DNA: pPV33, pPV501, recombinant plasmids from His$^+$ (pHIS) and Ara$^+$ (pARA) clones.

Bacteriophage lambda DNA: Molecular weight size standard, prepared in Exercise 6.

Note: In previous exercises, recombinant plasmids containing the *his* gene (pHIS) and *ara* genes (pARA) have been constructed, phenotypically characterized, and circular restriction maps of the cloned DNA fragments generated. In this exercise, the promoters of these *his* and *ara* genes will be cloned in the promoter-probe plasmids pPV501 and pPV33, respectively. Both vectors (see Appendix 3 for restriction maps) utilize the Ap^r gene of pBR322 as the primary selective marker. In pPV501 insertion of promoter-active DNA fragments in the unique *Bam*HI site will activate a silent Cm^r gene, while the insertion of promoter-active DNA fragments in the unique *Eco*RI site of pPV33 will activate the silent Tc^r gene of the vector. The *his* promoter can be obtained on a *Bgl*II fragment of pHIS, and the *ara* promoter on an *Eco*RI* fragment of pARA. This exercise typically requires 6 to 8 days for completion and utilizes many of the methods introduced in previous exercises.

Protocol

DAY 1

1. Use the large-scale plasmid purification procedure outlined in Exercise 4 to isolate plasmid DNA from clones containing pHIS or pARA recombinant plasmids. (Miniscreen DNA may also be used to clone the *his* operon and *ara* promoters.) Start a 20 ml LB culture of *E. coli* RRI.

2. Complete large-scale plasmid purification. Prepare competent RRI cells according to Exercise 9.

EXERCISE 14

CLONING THE PROMOTERS
FOR THE ARABINOSE AND
HISTIDINE OPERONS OF *E. COLI*

DAY 2

1. Using the plasmid DNA made from the pHIS and pPV501 DNA made in Exercise 4, set up the following restriction digests:

Reaction 9–1a	μl	Reaction 9–1b	μl
3 μg pPV501 DNA	6	3 μg pHIS DNA	6
3 units *Bam*HI endo.	3	3 units *Bgl*II endo.	3
10X *Bam*HI Buffer	3	10X *Bgl*II Buffer	3
ddH$_2$O	18	ddH$_2$O	18
	30		30

2. Using the plasmid DNA made from the pARA and pPV33 DNA made in Exercise 4, set up the following restriction digests:

Reaction 9–2a	μl	Reaction 9–2b	μl
3 μg pPV33 DNA	6	3 μg pARA DNA	6
3 units *Eco*RI endo.	3	3 units *Eco*RI endo.	3
10X *Eco*RI Buffer	3	10X *Eco*RI* Buffer	3
ddH$_2$O	18	80% glycerol	7.5
	30	ddH$_2$O	10.5
			30

3. Except for reaction 9–2b, transfer all reaction tubes to 37°C and incubate for 1 hour. Incubate reaction 9–2b at 37°C for 4 to 6 hours. After incubation, take 10 μl from each reaction and add it to a microfuge tube with 5 μl of reaction stop mix.

4. Set up a 1% agarose submarine gel or vertical slab gel as shown in Figs. 5.4 and 5.5, Chapter 5. Load the reactions and volumes as indicated below:

—Gel—

Well No.:	1	2	3	4	5	6	7	8	9	10	11	12
Reaction:			9–1a	9–1b		9–2a	9–2b					
Molecular Weight Markers:					4–10							
μl loaded: (vert. slab)			10	10	10	10	10					
μl loaded: (minigel)			5	5	5	5	5					

5. After electrophoresis (see Chapter 5), stain the gel in ethidium bromide and irradiate with UV light to determine the progress of the endonuclease digestion reactions. When it is established that the plasmid DNA has been digested to completion, add reactions 9–1a to 9–1b and 9–2a to 9–2b. To each combined reaction, add 100 μl of

buffer-saturated phenol, 100 µl of chloroform, 5 µl of 2 M NaCl, and 70 µl of TEN buffer. Vortex 30 seconds. Separate the phases by centrifugation in a microfuge at 4°C for 10 minutes. Remove 80 µl from the aqueous (upper) phase. Add 100 µl of cold isopropanol to precipitate the DNA. Collect the precipitated DNA by centrifugation in a microfuge (15 minutes, 4°C). Carefully pour off the supernatant and add 1.5 ml of 70% EtOH to each tube. Invert tubes several times to wash DNA pellets and remove excess salt and phenol. Centrifuge the samples to pellet the DNA, carefully pour off the ethanol supernatant, and dry the DNA pellets in a vacuum (5 to 15 minutes). Resuspend each DNA pellet in 20 µl ddH$_2$O.

6. Now that the reactions have been terminated and the restriction enzymes removed by phenol/chloroform extraction, set up the following ligation reactions in 1.5 ml microfuge tubes. If miniscreen DNA was used in these digestions, the concentration of DNA will be less than if purified DNA was used. Therefore, the ligation reaction volumes should be reduced to a total of 20 µl.

Reaction 9-3	µl	Reaction 9-4	µl
BglII cut pHIS and BamHI cut pPV501	20	EcoRI* cut pARA and EcoRI cut pPV33	20
1 unit T$_4$ ligase	1	1 unit T$_4$ ligase	1
10X T$_4$ ligase buffer	5	10X T$_4$ ligase buffer	5
10X ATP	5	10X ATP	5
ddH$_2$O	19	ddH$_2$O	19
	50		50

7. Prior to the addition of ligase, remove a 2 µl sample from each ligation reaction, add it to 8 µl Reaction Stop Mix (designate these samples 9-3a and 9-4a), and store the samples in the refrigerator until the ligation reaction is complete. These samples are controls and will be compared to your sample reactions to determine the degree of ligation.

8. Incubate the reaction mixtures at 12°C for 12 to 18 hours.

DAY 3

1. At the end of the incubation period, remove 2 µl of the ligated DNA and add it to 8 µl of Reaction Stop Mix. Designate the pHIS/pPV501 ligation reaction sample 9-3b and the pARA/pPV33 ligation reaction sample 9-4b. Set up a minigel or vertical slab gel. Load the samples as indicated below:

—Gel—

Well No.:	1	2	3	4	5	6	7	8	9	10	11	12
Reaction:		9-3a	9-3b	4-10	9-4a	9-4b						
µl loaded: (vert. slab)		10	10	10	10	10						

EXERCISE 14

CLONING THE PROMOTERS FOR THE ARABINOSE AND HISTIDINE OPERONS OF E. COLI

EXERCISE 14

CLONING THE PROMOTERS FOR THE ARABINOSE AND HISTIDINE OPERONS OF *E. COLI*

2. When the ligation reaction is complete, terminate the ligation reaction by incubating the samples at 65°C for 5 minutes. Sterilize the reaction by centrifugation for 2 minutes in the microfuge, transfer the supernatant to a sterile 1 cm × 7.5 cm glass test tube, and add 10 µl of sterile 0.3 M $CaCl_2$.

3. Transform competent RRI cells with the ligated DNA preparations according to the procedure outlined in Exercise 9.

4. Spread 0.1 ml of each transformation mixture on each of 5 LB-Ap plates and incubate at 37°C overnight.

DAY 4

1. Using the scoring template provided in the back cover, patch as many transformants as possible on the selective media indicated below:

Transformation Plate	Test Plates	Master Plate
His promoter transformants	LB-Cm (10 µg/ml)	
LB-Ap	LB-Tc (10 µg/ml)	LB-Ap (20 µg/ml)
Ara promoter transformants	LB-Cm (10 µg/ml)	
LB-Ap	LB-Tc (10 µg/ml)	LB-Ap (20 µg/ml)

2. Patch the transformants according to the procedure outlined in Exercise 10.

3. After completing the transfers, incubate the test and master plates at 37°C to allow for colony formation. Colonies should appear within 24 hours on LB agar plates.

DAY 5

1. Examine the drug-resistance phenotypes of the colonies and streak the putative pPHIS (*his* promoter plasmids) (Ap^r, Tc^s, Cm^r) and the putative pPARA (*ara* promoter plasmids) (Ap^r, Tc^r, Cm^s) transformants on sectored LB-Ap plates and incubate the plates at 37°C overnight.

DAY 6

1. Pick single colonies from each of the putative promoter plasmid clones and inoculate 1 ml of LB. Incubate the cultures overnight at 37°C.

DAY 7

1. Following the miniscreen procedure outlined in Exercise 5, isolate plasmid DNA from each of the cultures.

2. To test whether the promoters were cloned into pPV501 or pPV33, set up the following restriction digests:

Reaction 9-5a	μl		Reaction 9-5b	μl
0.5 μg pPV501	1		miniscreen DNA	3
1 unit EcoRI endo.	1		1 unit EcoRI endo.	1
10X EcoRI Buffer	2		10X EcoRI Buffer	2
ddH₂O	16		ddH₂O	14
	20			20

Reaction 9-6a	μl		Reaction 9-6b	μl
0.5 μg pPV33	1		miniscreen DNA	3
1 unit EcoRI endo.	1		1 unit EcoRI endo.	1
10X EcoRI Buffer	2		10X EcoRI Buffer	2
ddH₂O	16		10X EcoRI Buffer	14
	20			20

EXERCISE 14

CLONING THE PROMOTERS FOR THE ARABINOSE AND HISTIDINE OPERONS OF *E. COLI*

3. If the *his* operon promoter has been cloned, the smallest pPV501 fragment (approximately 850 bp) will not be observed. Instead of the 850 bp fragment, a band will be observed migrating at approximately 2,050 bp. If the *ara* operon promoters have been cloned, two bands will be observed on the gel. One will be equivalent in size to the pPV33 vector band and the second will be approximately 2,500 bp.

It has been shown (Greenfield et al., 1978; Frunzio et al., 1981) that the restriction fragments carrying the *his* and *ara* regulatory sequences contain bidirectional promoters. How would you determine which promoter is acting on the marker genes in pPV33 and pPV501?

EXERCISE 15

Using Antibiotic Resistance as an Indirect Measure of Promoter Strength

Materials

M9 Minimal Media (Tc-M9 min.-gluc): 6 plates supplemented with 0.2% glucose, 42 μg/ml histidine, 166 μg/ml proline, 0.2 μg/ml thiamine hydrochloride (B_1). One plate each should contain tetracycline (Tc) at a concentration of 5, 10, 20, 40, 60, and 80 μg/ml.

M9 Minimal Media (Tc-M9 min.-ara): 6 plates supplemented with 0.2% arabinose, 42 μg/ml histidine, 166 μg/ml proline, 0.2 μg/ml thiamine hydrochloride (B_1). One plate each should contain Tc at a concentration of 5, 10, 20, 40, 60, and 80 μg/ml.

M9 Minimal Media (Cm-M9 min.-gluc): 6 plates supplemented with 0.2% glucose, 166 μg/ml proline, 41 μg/ml leucine, 54 μg/ml adenine, and 0.2 μg/ml thiamine hydrochloride (B_1). One plate each should contain chloramphenicol (Cm) at a concentration of 25, 50, 75, 100, 125, and 150 μg/ml.

M9 Minimal Media (Cm-M9 min.-gluc + amitrole): 6 plates supplemented with 0.2% glucose, 166 μg/ml proline, 41 μg/ml leucine, 0.2 μg/ml thiamine hydrochloride (B_1), 54 μg/ml adenine, and 1.7 mg/ml amitrole (3-amino-1,2,4-triazole, Sigma—*Caution:* possibly carcinogenic). One plate each should contain Cm at a concentration of 25, 50, 75, 100, 125, and 150 μg/ml.

Luria Broth (LB): 1 small test tube containing 3 ml.

7 agar plates (1.5%) each containing 20 μg/ml ampicillin.

Competent cells: BE42 cells prepared according to Exercise 9.

Note: In Exercise 14, several *his* and *ara* promoter-active recombinants should have been identified on the basis of phenotypic and restriction enzyme analysis. In this exercise these promoter-active recombinants will be further examined with regard to promoter strength and regulatory properties. The results obtained from these experiments should enable you to determine which promoter (e.g., *ara* B or *ara* C) is responsible for expression of the silent markers on pPV33 and pPV501.

Protocol

DAY 1

1. Using the pPHIS/RRI promoter-active clones identified in Exercise 14, patch cells onto the following series of minimal agar plates:

   pPHIS/RRI Clones
   - Cm-M9 min.-gluc: 25, 50, 75, 100, 125, 150 μg/ml Cm
   - Cm-M9 min.-gluc + amitrole: 25, 50, 75, 100, 125, 150 μg/ml Cm

   (After the sixth plate containing 150 μg/ml Cm, patch cells onto an LB-Ap master plate.)

2. Incubate the patched minimal agar plates from 18 to 24 hours at 37°C.

3. Using miniscreen DNA of *ara* promoter-active recombinants prepared in Exercise 14 (pPARA), transform competent cells of BE42, prepared according to Exercise 9, (since RRI is Ara$^-$, the BE42 strain must be used for this part of the exercise).

4. Dilute the transformation mixture into 3 ml of LB and incubate at 37°C (with aeration) for 1 to 3 hours. Plate 0.1 ml on each of 3 LB-Ap agar plates. Incubate plates at 37°C overnight.

DAY 2

1. Using the pPARA/BE42 promoter-active clones present on the LB-Ap plates prepared on Day 1, patch cells onto the following series of minimal agar plates:

   pPARA/BE42 Clones
   - Tc-M9 min.-gluc: 5, 10, 20, 40, 60, 80 μg/ml Tc
   - Tc-M9 min.-ara: 5, 10, 20, 40, 60, 80 μg/ml Tc

   (After the sixth plate containing 80 μg/ml Tc, patch cells onto an LB-Ap master plate.)

2. Incubate the patched minimal plates from 18 to 24 hours at 37°C.

DAY 3

1. Record the results obtained from patching the Cm-M9 min. and Tc-M9 min. plates. Did any of the clones grow to a higher level of drug resistance on the arabinose or amitrole-containing media? Can these results be used to determine which promoter is acting on the marker genes with pPV33 and pPV501?

If further quantitation of the drug resistance levels is desired, EOP$_{50}$ determination of the pPARA clones can be performed as described in the introductory section to this chapter. Using the protocol for determining CAT activity provided in Appendix 2, the levels of Cm^r expressed by the pPHIS clones can also be determined.

APPENDIX 1

Recipes

LB Medium

a. Bacto tryptone   10 gm
    Bacto yeast extract   5 gm
    NaCl   10 gm
    dH_2O to   1 liter volume

    Adjust pH to 7.0 with 5 N NaOH (approximately 0.4 ml/liter)

b. For agar plates, add
    Bacto agar   15 gm/liter

    After autoclaving, cool medium to 45°C before adding antibiotics

c. For stabs, add
    Bacto agar   8 gm/liter

    Boil to melt agar, dispense 2 ml aliquots into 1 dram vials, cap, and autoclave 5 min to sterilize.

YT Medium

a. Bacto tryptone   8 gm
    Bacto yeast extract   5 gm
    NaCl   5 gm
    dH_2O to   1 liter volume

b. For plates, add
    Agar   15 gm/liter

c. For soft agar overlay add .6 gm agar to 100 ml bottles of YT. Autoclave. Can microwave or boil to melt prior to each use.

APPENDIX 1
RECIPES

M9 Medium

Sterile 10X Salt*	100.0 ml	(Add after autoclaving other ingredients.)
20% glucose	20.0 ml	
.01 M $CaCl_2$	10.0 ml	
0.1 M $MgSO_4$	10.0 ml	
20% Casamino acids	20.0 ml	
H_2O	840.0 ml	

*Salt Mix	10X
Na_2HPO_4	70 gm
or ($Na_2HPO_4 \cdot 7 H_2O$)	(132 gm)
KH_2PO_4	30 gm
NaCl	5 gm
NH_4Cl	10 gm
H_2O	1000 ml

For M9 minimal medium, omit casamino acids. Other carbon sources may be substituted for glucose. Supplementation levels for amino acid auxotrophs can be found in the accompanying chart. Add proline, leucine and B1 for the RR1 strain.

Supplementation Levels for Amino Acids

Amino Acid	Final concentration µg/ml	Stock	Amount per liter
L-alanine	75	1%	7.5 ml
L-arginine	145	2%	7.25 ml
L-asparagine	75	1%	7.5 ml
L-aspartic acid	75	1%	7.5 ml
L-cysteine	50	2%	2.5 ml
L-histidine	42	2%	2.1 ml
L-glutamic acid	75	1%	7.5 ml
L-glutamine	75	1%	7.5 ml
glycine	110	2%	5.5 ml
L-isoleucine	42	1%	4.2 ml
L-leucine	41	1%	4.1 ml
L-lysine	75	1%	7.5 ml
L-methionine	25	2%	1.25 ml
L-phenylalanine	75	1%	7.5 ml
L-proline	164	4%	4.1 ml
L-serine	42	2%	2.1 ml
DL-threonine	82	2%	4.1 ml
L-tryptophan	18	0.25%	7.1 ml
L-tyrosine	75	1%	7.5 ml
L-valine	42	1%	4.2 ml
Vitamin B1, (thiamine HCl)	.166	0.1%	0.166 ml

Freezing Medium for Bacterial Strains

APPENDIX 1
RECIPES

2X Stock Solution:
1% Yeast Extract
10% Dimethylsulfoxide
10% Glycerol
0.2 M K_2HPO_4/NaH_2PO_4 pH = 7.0

Filter sterilize this stock solution and store refrigerated or frozen. Bacterial strains are grown in their normal media to early stationary phase and diluted with an equal volume of the above 2X stock solution. The 1/2 dram vials may be conveniently filled with a syringe and needle (1/2 ml 2X media and 1/2 ml cells). Strains may be stored frozen at $-20°$ or preferably $-70°C$. For organisms that settle out of solution quickly it is preferable to quick freeze in liquid nitrogen or dry ice ethanol. Frozen samples of cells are removed with a sterile wooden applicator stick.

Yeast Minimal Medium*

10X salts	100 ml	
20% glucose	100 ml	(Add after autoclaving
1000X metals	1.0 ml	other ingredients.)
1000X vitamins	1.0 ml	
100X supplements	10 ml	
ddH_2O to	1 liter volume	
pH to 6.0–6.5		
Agar (for solid medium)	20 gm	

10X yeast salts (per liter)

$CaCl_2 \cdot 2H_2O$	1 gm
NaCl	1 gm
$MgSO_4 \cdot 7H_2O$	5 gm
KH_2PO_4	10 gm
$(NH_4)_2SO_4$	50 gm

Trace Metals 1000X Stock Solution

Boric acid	50 mg
Copper sulfate $\cdot 5H_2O$	4 mg
Potassium iodide	10 mg
Ferric chloride $\cdot 6H_2O$	20 mg
Manganese sulfate $\cdot H_2O$	40 mg
Sodium molybdate $\cdot 2H_2O$	20 mg
Zinc sulfate $\cdot 7H_2O$	40 mg
ddH_2O to	100 ml

Dissolve chemicals in order listed in 100 ml ddH_2O. There will be a precipitate which will not dissolve. Shake bottle before using.

---
*Recipes provided courtesy of Dr. S. R. Snow, Department of Genetics, University of California, Davis.

APPENDIX 1
RECIPES

Vitamins 1000X Stock Solution

Biotin	0.1 mg
Calcium pantothenate	20 mg
Folic acid	0.1 mg
Inositol	100 mg
Niacin	20 mg
P-Aminobenzoic acid	10 mg
Pyridoxine HCl	20 mg
Riboflavin	10 mg
Thiamine HCl	20 mg

Put indicated amounts into 100 ml flask, add 50 ml deionized water to dissolve. (Riboflavin will not entirely dissolve at room temperature.) Make 5 ml aliquots and store frozen.

Supplements for Minimal Medium (100X Stock Solutions)

Amino acids	mg/ml
L-alanine	2
L-arginine	2
L-asparagine	2
L-aspartic acid	10
L-cysteine	2
glycine	2
L-glutamine	2
L-glutamic acid	10
L-histidine (free base)	2
L-isoleucine	3
L-leucine	3
L-lysine	3
L-methionine	2
L-phenylalanine	5
L-proline	2
L-serine	37.5
L-threonine	20
L-tryptophane	2
L-tyrosine	3
L-valine	10

Bases

adenine sulfate	2
uracil	1

Other supplements

cyclohexamide (actidione)	1.4
canavanine H_2SO_4	6

YEPD (Yeast) Medium

Difco peptone	20 gm
Difco yeast extract	10 gm
20% Dextrose	100 ml (Add after autoclaving other ingredients.)
ddH$_2$O to	1 liter
Agar (for solid medium)	20 gm

APPENDIX 1
RECIPES

This medium will support the growth of most auxotrophic strains without additional supplements, and is used as a general growth medium. For storage slants, the amount of each ingredient (except for the agar) is doubled.

YNBD (Yeast Minimal) Medium

Difco yeast nitrogen base without amino acids	7 gm
Dextrose	2 gm
100X supplements	10 ml
ddH$_2$O to	1 liter
Agar (for solid medium)	20 gm

Add the appropriate 100X supplement depending on the auxotrophic markers of the yeast strain.

10X Tris-Borate Buffer

Trizma base	216 gm
EDTA (disodium salt)	18.6 gm
Boric acid	110 gm
ddH$_2$O to 2 liter volume	
pH should be 8.0–8.2	

Phenol Equilibration Buffer

50 mM Tris-HCl pH 7.5
100 mM NaCl

Add equal volume to phenol. Stir overnight at 4°C.

A50 Buffer/Dowex Buffer

(50 mM)	Trizma base	24 gm
(500 mM)	NaCl	117 gm
(1 mM)	0.25 M EDTA	16 ml
(1 mM)	1 M NaN$_3$	4 ml
	ddH$_2$O	3.5 liters

Adjust pH to 8.0 with 8–10 ml concentrated HCl
ddH$_2$O to   4 liters

APPENDIX 1
RECIPES

TEN (DNA) Buffer

10 mM Tris-HCl pH 7.6 (10 ml, 1.0 M stock)

1 mM EDTA (10 ml, 0.1 M stock)

10 mM NaCl (2 ml, 5.0 M stock)

ddH$_2$O to 1 liter volume

10X SSC (Standard Saline-Citrate)

1.5 M NaCl	87 gm
0.15 M NaCitrate	44 gm
ddH$_2$O to	1 liter volume

Dowex Preparation

Materials

Dowex AG 50W-X8:   100–200 mesh (BioRad).

ddH$_2$O.

NaOH: Solid.

HCl: Concentrated.

50 mM Tris-HCl pH 8.0.

Dowex Buffer: Also A50 buffer.

Protocol

1. Place 100 ml Dowex AG 50W-X8 (resin) and 400 ml ddH$_2$O in a 500 ml beaker.
2. Add 40 gm solid NaOH (2 N) and stir until dissolved.
3. Let the Dowex settle and decant the supernatant.
4. Wash 3 times with ddH$_2$O.
5. Bring volume up to 500 ml with ddH$_2$O.
6. Slowly add 10 ml concentrated HCl with constant stirring. Stir for 2 to 5 minutes.
7. Decant the supernatant and wash the resin with one liter ddH$_2$O.
8. Repeat steps 5 through 9.
9. Wash 5 times with ddH$_2$O.
10. Bring pH to 8.0 with 50 mM Tris-HCl pH 8.0, allowing time for equilibration.
11. Decant the Tris-HCl buffer and resuspend in 2 volumes of Dowex Buffer. Store at 4°C.
12. Resin is typically used to remove ethidium bromide as described in the protocols for "Plasmid Isolation and Purification" and "Extraction of Restriction Fragments from Low-Melt Agarose Gels."

APPENDIX 2

Protocols

Plasmid DNA Isolation Using the CsCl Dye-Buoyant Density Gradient Method

There are many methods currently being used to isolate plasmid DNA in a highly pure state. All of these techniques are designed to separate chromosomal and open circle plasmid DNA from supercoiled plasmid DNA. Small relaxed plasmids (capable of DNA replication in the absence of protein synthesis) such as pBR322 are particularly amenable to these techniques because they can be amplified to nearly 50% of the total cellular DNA with antibiotics such as chloramphenicol (Clewell, D. B., J. Bacteriol. 110:667, 1972; Bolivar et al., Gene 2:75, 1977). The procedure described below is not necessarily the easiest or quickest, but it has been proven to yield plasmid DNA free of contaminating oligonucleotides and chromosomal DNA.

Materials

SET Buffer: 20% sucrose, 50 mM Tris-HCl pH 7.6, 50 mM EDTA.

M9 Medium: 1 liter, supplemented with casamino acids and vitamin B_1.

Lysozyme: 5 mg/ml in TEN Buffer.

TEN Buffer: 10 mM Tris-HCl pH 7.6, 1 mM EDTA, 10 mM NaCl.

RNase Buffer: Pancreatic ribonuclease A, 10 mg/ml in 0.1 M Na acetate, 0.3 mM EDTA, pH 4.8, preheated to 80°C for 10 minutes.

APPENDIX 2

PROTOCOLS

3X Triton Lytic Mix:
- 3 ml 10% Triton X-100
- 75 ml 0.25 M EDTA (Disodium salt)
- 15 ml 1 M Tris-HCl pH 8.0
- <u>7 ml</u> ddH$_2$O
- 100 ml final volume.

Phenol: Buffer equilibrated (see Phenol Equilibration Buffer in Appendix 1).

CsCl: 2.2 gm/gradient.

Ethanol: 95%.

A50 Buffer: See Appendix 1.

Dowex 50W-X8: 100–200 mesh. J. T. Baker or Biorad. Precycled through acid and base.

Propidium Iodide (PdI): 2 mg/ml in ddH$_2$O.

Mineral Oil.

Dry Ice.

Chloramphenicol.

Chloroform.

5 M NaCl.

Protocol

1. Add 2 ml of starter culture to 1 liter of M-9 media supplemented with glucose and casamino acids. Add vitamins and other supplements as required.

2. Incubate the culture at 37°C with aeration to a cell density of 4–5 × 10^8 cells/ml (140 Klett units—green filters; A$_{450}$ = 1.01). Add chloramphenicol (190–200 µg/ml) and continue incubation for 16 hours. This step should only be used for relaxed plasmids. It is important to add chloramphenicol well before the culture goes into stationary phase.

3. At this point cell cultures may be stored in the cold for several hours if needed. Centrifuge the culture to pellet the cells using Sorvall GSA or HS-4 rotor, 6.5 krpm, 20 minutes, 4°C.

4. Freeze cell pellets in a dry ice/EtOH bath for about 10 minutes or at −70°C or −20°C for 5 hours. This step weakens the cell wall and prepares the cells for the lysozyme treatment.

5. Suspend cell pellets in a total of 10 ml of SET Buffer and transfer to 50 ml polypropylene centrifuge tubes. Keep on ice. Sucrose helps maintain the integrity of the weakened cell wall. Clumps of cells should be separated by mixing, but care must be taken not to lyse cells at this time.

6. Add the following solutions per liter of cells: 1 ml lysozyme and 0.1 ml RNase Buffer. (Stock solutions of lysozyme and RNase should be stored frozen in 2–5 ml aliquots.)

APPENDIX 2
PROTOCOLS

7. Mix gently and let stand on ice for 15 minutes. (The cells are now converted to spheroplasts.)
8. Add 5 ml of 3x Triton Lytic Mix. Mix gently and let stand on ice for another 15 minutes. (Spheroplasts are lysed, plasmids released, and the major portion of the bacterial chromosomal DNA remains associated with the membrane.)
9. Centrifuge the lysis mixture in 50 ml polypropylene tubes in the Sorvall SS34 rotor at 17 krpm for 40 minutes, 4°C. This pellets chromosome-membrane complexes and leaves plasmid DNA in the supernatant.
10. Decant supernatant into a plastic graduated cylinder (DNA sticks to glass) immediately after removing the tubes from rotor. Note the volume and pour into a 250 ml plastic centrifuge bottle. Add 2/3 volume ddH$_2$O (makes sucrose less dense than phenol).
11. Add 2/3 of total volume of cold buffer-saturated phenol to solution and mix thoroughly. Centrifuge in the HS-4, swinging bucket rotor at 6.5 krpm, 10 minutes, 4°C. (Wear gloves and avoid spilled phenol.)
12. Add a volume of chloroform equal to the phenol. Mix well and centrifuge in the HS-4 rotor, 6.5 krpm, 15 minutes, 4°C. Remove upper (aqueous) phase and place into a 250 ml bottle. (Avoid protein pellicle at interface.) The chloroform step facilitates the formation of a firm pellicle and distinct phases.
13. Add 1/25 volume of 5 M NaCl to aqueous phase. Then add two volumes of −20°C EtOH. Mix thoroughly and let stand at −20°C for one hour or longer to precipitate the DNA. Centrifuge at 6.5 krpm, 60 minutes, −5°C. Discard supernatant and dry excess liquid from bottle with kimwipes.
14. Resuspend pellet in 3 ml A50 buffer.* Load suspension onto a 2 × 30 cm A50 (Bio-Rad) column equilibrated with A50 buffer. Collect 80, 2 ml fractions. If you wish to load the column by displacement (adding DNA solution under the buffer-head), add 1 ml of 80% glycerol to 2 ml of DNA-A50 buffer solution. A50 chromatography removes all low molecular weight impurities (e.g., amino acids, polypeptides, oligonucleotides, phenol) (see fig. 3.2 in Chapter 3).
15. Determine A$_{260}$ of each fraction. DNA should be in the first 40 tubes, usually between tubes 15–25. An A$_{260}$ of 1.0 = 50μg of DNA/ml. Pool DNA peak. Add 2 volumes of −20°C EtOH, and store at −20°C for 1 hour or more. Centrifuge in HS-4 rotor, 6.5 krpm, 60 minutes, −5°C. Resuspend DNA pellet (∼500 μg–1,000 μg) in 4.3 ml of TEN Buffer.

CsCl/PdI Gradient Centrifugation: Propidium iodide "PdI" intercalates double-stranded linear or nicked-circular DNA to a greater extent than supercoil DNA. Since PdI is less dense than DNA, nonsupercoiled DNA assumes a lower buoyant density than

---
*Some laboratories resuspend the DNA in 4 ml of TEN Buffer and proceed to Step 16, circumventing the A50 column chromatography. However, DNA prepared in this manner can be contaminated with oligonucleotides.

supercoiled DNA. Without PdI the superhelical and nonsupercoiled forms of DNA would form two narrowly separated bands in the CsCl gradient. While ethidium bromide can increase the separation of the two DNA forms, PdI gives a greater degree of separation and therefore facilitates fractionation.

16. Preparation of CsCl gradients: Into 5 ml cellulose nitrate tubes (½″ × 2″) add 2.2 g CsCl and 2.1 ml DNA in TEN (each tube should contain no more than 500 µg DNA). Invert the tubes several times to dissolve the CsCl.

    Caution: Work from this point should be done in low light or with safelight, as visible light may cause PdI to fluoresce and nick supercoiled DNA.

17. Next add 150 µl PdI and mix with DNA-CsCl, using a piece of parafilm. Layer 2.4 ml of mineral oil on top of the gradient to keep the sides of the tube from collapsing. All buckets should be balanced with mineral oil. Do not spill oil on outer surface of centrifuge tubes.

18. Centrifuge in the Beckman SW50.1 rotor, 36 krpm for 20 hours at 20°C. The DNA in the gradient can be visualized with a UV mineral light. Two bands should be visible. The upper band is bacterial and nicked plasmid DNA while the lower band is covalently closed plasmid DNA (replicative intermediates of plasmid produce a smear between bands).

19. Using a UV illuminator, pierce the bottom of the tube and collect plasmid DNA drop-wise in a plastic tube.

20. Remove PdI from the DNA with Dowex 50W-X8 (100–200 mesh). About 1½ ml bed volume per gradient tube is placed into a small column and washed with 10 ml of A50 Buffer/ml of Dowex resin. Dilute DNA with an equal volume of A50 Buffer and apply the sample to the Dowex column. If the DNA is not diluted, the sample may pass through the column too rapidly to allow complete removal of the PdI.

21. Apply the DNA to the column and elute slowly into plastic tubes. Add at least one bed volume of A50 Buffer to wash out any remaining DNA. Check with UV mineral light to see if the PdI has been removed. If not, remove the top part of the Dowex column (that which is stained red) and run the DNA over the column again. Beyond this point, work can be done in full light.

22. Dialyze the DNA against 4 liters of 10 mM Tris, 1 mM EDTA pH 8.0 for 12 hours or more at 4°C. Remove DNA from dialysis tube, and precipitate by adding 1/25 volume 5M NaCl and 2 volumes −20°C EtOH. Let sit 1 hour or longer. Centrifuge in HS-4 rotor, 6.5 krpm, 60 minutes at −5°C.

23. Resuspend the DNA in 1–2 ml TEN Buffer. Check A_{260} to determine DNA concentration (0.8–1.5 mg of DNA/liter of culture is common). Store at −20°C or in the refrigerator over a drop of chloroform.

Isolation of High Molecular Weight Plasmid DNA*

APPENDIX 2
PROTOCOLS

Material

TNS Buffer: 10 mM Tris-HCl pH 8.0, 100 mM NaCl, 20% sucrose.

Lysis Solution: 100 mM Tris-HCl pH 8.0, 50 mM EDTA, 0.5 M NaCl, 2% SLS (sodium lauryl sarcosinate; Giegy).

TE Buffer: 10 mM Tris-HCl pH 7.6, 1 mM EDTA (disodium salt).

Lysozyme: 10 mg/ml in TE Buffer.

RNase Buffer: Pancreatic ribonuclease A, 10 mg/ml, in 0.1 M Na acetate, 0.3 mM EDTA, pH 4.8, preheated to 80°C for 10 minutes.

0.5 M EDTA (disodium salt) pH 8.0.

3 M NaOH.

2 M Tris-HCl pH 5.0.

NaCl: Solid.

Polyethylene Glycol 6000: Solid.

Phenol/Chloroform: 1 to 1 mixture (v/v).

Chloroform/n-amyl Alcohol: 24 to 1 mixture (v/v).

CsCl: Solid.

Ethidium Bromide: 4 µg/ml made up in ddH$_2$O.

Butanol: Saturated with ddH$_2$O.

Dialysis Tubing.

Protocol

DAY 1

1. Harvest a 500 ml culture of bacteria by centrifugation (Sorval GSA rotor, 8 krpm, 4°C, 20 minutes). Freeze the pellet at −20°C and thaw at room temperature.

2. Resuspend the cell pellet in 100 ml of TNS Buffer. Add 40 ml of lysozyme and 3 ml of RNase Buffer. Incubate on ice for 30 minutes.

3. Add 120 ml of Lysis solution and mix thoroughly. Incubate at 25°C until the solution clears.

4. Adjust the pH of the solution to 12.1 by the dropwise addition of 3 M NaOH, with gentle stirring.

5. After the pH has equilibrated to 12.1, adjust the pH to 8.5 by slowly adding 2 M Tris-HCl pH 5.0, with stirring.

6. Measure the volume and add 5.8 gm of solid NaCl per 100 ml of solution to give a final concentration of 1 M NaCl. Incubate on ice for 1 hour.

7. Centrifuge as in Step 1. Measure the volume of the supernatant and add solid PEG 6000 to a final concentration of 5% (w/v). Carefully dissolve PEG and store overnight at 4°C.

---

*Kao, Perry, and Kado (1982). Mol. Gen. Genet. 188, 425. Modifications provided by Dr. R. C. Lundquist, Department of Plant Pathology, University of California, Davis.

APPENDIX 2
PROTOCOLS

DAY 2

1. Collect the resulting precipitate by centrifugation in a Sorvall GSA rotor, 10 krpm, 4°C, for 30 minutes. Pour off and discard the supernatant.
2. Resuspend the pellet in 10 ml of TE, transfer to a 30 ml Corex tube, and extract three times with an equal volume of phenol/chloroform.
3. Extract the aqueous phase three times with an equal volume of chloroform/n-amyl alcohol.
4. Combine 8 ml of the resulting aqueous phase with 8.15 gm CsCl and 0.62 ml ethidium bromide. Adjust density to 1.638 gm/ml.
5. Centrifuge in a Sorvall TV-865 rotor, 50 krpm, 20°C, for 15 hours to resolve the supercoiled plasmid DNA from contaminants. Note that the CsCl density of 1.638 gm/ml has been determined for the large plasmids present in *A. tumefaciens*. It may be necessary to adjust the density or the centrifugation speed to obtain optimal separation of other plasmids.

DAY 3

1. Use a long-wave ultraviolet light source to visualize the plasmid DNA (lower band). Use a syringe fitted with an 18 gauge needle to withdraw the DNA band. Puncture the side of the tube below the band, tilt the needle upward into the DNA, and withdraw the band.
2. Add the band to a second CsCl-ethidium bromide gradient prepared as in Step 4 of Day 2. Centrifuge in a Sorvall TV-865 rotor, 50 krpm, 20°C, for 8 hours.
3. Withdraw the purified plasmid DNA as in Step 1 of Day 3.
4. Remove ethidium bromide by extraction of the plasmid DNA with water-saturated n-butanol.
5. Remove CsCl by 24 hour dialysis against 4 liters TE Buffer.
6. Store DNA at 4°C.

Miniscreen for Large Plasmids*

This miniscreen has proven useful with plasmids up to 300 kb in size.

Materials

Lysis Solution:

Tris base	1.21 gm.	(50 mM)
SDS (sodium dodecyl sulfate, BDH Chemicals Ltd.)	6.0 gm.	(3%)
2 N NaOH, *fresh*	8.2 ml	
ddH$_2$O to	200 ml volume.	

*Kado and Liu (1981). J. Bacteriol. 145, 1365. Protocol provided courtesy Dr. C. I. Kado, Department of Plant Pathology, University of California, Davis.

APPENDIX 2
PROTOCOLS

Phenol/chloroform (1:1, v/v): The phenol should be freshly distilled and stored frozen until use. Once mixed with chloroform, the solution can be stored at room temp.

E Buffer 20X Stock:

0.8 M Tris base

0.1 M Na acetate

0.02 M Na_2EDTA

Adjust pH to 7.9 with glacial acetic acid. Dilute 20-fold prior to use.

523 Medium:

Sucrose	10.0 gm
Casein hydrolysate	8.0 gm
Yeast extract	4.0 gm
K_2HPO_4	2.0 gm
$MgSO_4$	0.15 gm
ddH_2O to	1 liter volume.

Adjust pH to 7.1.

BCP Dye:

0.25% Bromocresol purple

50% glycerol

50 mM Tris-acetate pH 7.9

Protocol

DAY 1

1. Grow 5 ml bacterial cultures in 523 medium with shaking.

   Note: This protocol was designed for use with *A. tumefaciens*, and other media may be preferable for other bacteria.

DAY 2

1. Place 0.2–0.5 ml of culture in a microfuge tube and centrifuge 1 minute. The amount of culture used is dependent on the cell density: Use 0.2 ml for a cell density of 10^9 cells/ml, and 0.5 ml for a density of 10^8 cells/ml. Too many cells will result in poor lysis and decrease the yield of plasmid DNA.

2. Resuspend cell pellet in 100 µl of E Buffer by vigorous vortexing.

3. Add 200–300 µl of Lysis solution and invert the tubes 10 times.

4. Incubate at 55°C for 90 minutes.

5. Add 400 µl of phenol/chloroform, invert the tubes 100 times.

6. Centrifuge at room temperature for 3 minutes.

7. A distinct interface should form. If not, return to Step 1 of Day 2, reduce the amount of cells, and repeat the lysis procedure.

8. Without disturbing the interface, remove the upper aqueous phase. These DNA samples can be stored at 4°C for several days prior to electrophoretic analysis.

9. For electrophoretic analysis, prepare a 0.7% agarose gel made in 1X E Buffer. For plasmids greater than 100 kb, a gel 0.5 cm thick is desirable.

10. Mix 50 µl of plasmid DNA sample with 5 µl of BCP dye on a square of parafilm M. Avoid excess pipetting of the samples to minimize shear.

11. Load as much sample on the gel as is possible.

12. Electrophorese for 3 hours at 12 volts/cm gel length.

13. Stain the gel with ethidium bromide, destain with ddH$_2$O, and photograph.

Isolation of *Bacillus subtilis* Chromosomal DNA*

Materials

Brain Heart Infusion Broth (BHI): 1 liter

SET Buffer: 20% sucrose, 50 mM Tris-HCl pH 7.6, 50 mM EDTA.

RNase Buffer: Pancreatic ribonuclease A, 10 mg/ml, in 0.1 M Na acetate, 0.3 mM EDTA, pH 4.8, preheated to 80°C for 10 minutes.

TEN Buffer: 10 mM Tris-HCl pH 7.6, 1 mM EDTA, 10 mM NaCl.

Chloroform/n-amyl Alcohol: 24 to 1 mixture.

Lysozyme: 5 mg/ml in TEN Buffer.

Pronase: 2 mg/ml in TEN Buffer, preheated to 37°C for 15 minutes.

Sodium Dodecyl Sulfate (SDS): 25%.

NaCl: 5 M.

Ethanol: 95%, stored in a freezer.

Water (ddH$_2$O): Sterile, double distilled.

Dry Ice.

Phenol: Buffer saturated (see Appendix 1).

Protocol

DAY 1

1. Harvest 1 gm of cells (wet weight) from a BHI culture grown at 37°C with aeration, to log phase growth (approximately A$_{450}$ of 1.0).

2. To wash the cell pellet, resuspend in 20 ml of TEN Buffer (in a 50 ml polypropylene centrifuge tube) and spin at 6.5 krpm for 5 minutes at 4°C.

---

*Protocol courtesy of Dr. R. H. Doi, Department of Biochemistry, University of California, Davis.

3. Resuspend the cell pellet in 10 ml of SET Buffer and add 1 ml of lysozyme. Incubate at 37°C for 30 minutes. Monitor the formation of spheroplasts using a phase-contrast microscope.

4. Divide the cell suspension between 2 (sterile) 25 ml screw-cap Corex tubes. Add 5 ml of TEN Buffer and 0.5 ml of SDS. Tighten caps and gently invert tubes until the solution is thoroughly mixed and lysis occurs.

5. To each tube add 1 ml of 5 M NaCl and an equal volume (about 10 ml) of buffer saturated phenol. (Use plastic or rubber gloves when extracting protein with phenol.) Invert the tubes gently until the mixtures have been emulsified (about 5 minutes).

6. Separate phases by centrifugation at 6.5 krpm for 5 minutes at 4°C.

7. Using a wide-bore pipet (or the mouth-end of a 10 ml serological pipet), transfer the upper (aqueous) phase to fresh (sterile) 25 ml, screw-cap Corex tubes. Avoid protein pellicle at the interface.

8. Add an equal volume of chloroform/n-amyl alcohol and extract residual protein as described in Step 5. (Loosen the caps shortly after mixing to relieve pressure.)

9. After 5 minutes of mixing, separate phase as described in Step 6 and transfer upper phases from both tubes to a 100 ml glass beaker. Set on ice.

10. Precipitate the DNA by adding 2 volumes of cold 95% ethanol. Mix the solutions thoroughly and allow approximately 5 minutes for the DNA to precipitate.

11. Spool the DNA out of solution on a glass rod, dip into a tube of 95% ethanol, and resuspend in 10 ml of TEN Buffer (in a sterile 25 ml screw-cap Corex tube). To dissolve the precipitated DNA, leave overnight at 4°C.

DAY 2

1. Add 0.05 ml of RNase Buffer and incubate at 37°C for 2 hours. Next add 0.5 ml pronase and continue the incubation at 37°C for another hour.

2. Add 5 ml buffer saturated phenol, and 5 ml chloroform-n-amyl alcohol and mix gently for 5 minutes. Separate phase as in Step 6 of Day 1.

3. Repeat the extraction with 10 ml of chloroform-n-amyl alcohol only. Separate the phases as described in Step 6 of Day 1.

4. Transfer the upper phase to a 50 ml glass beaker and add 2 volumes of cold 95% ethanol. Mix and spool out DNA as described in Step 10 of Day 1. Rinse the precipitated DNA in 95% ethanol and resuspend in 2–4 ml of TEN Buffer.

5. Determine the A_{260}/A_{280} ratio and estimate the DNA concentration.

APPENDIX 2
PROTOCOLS

Rapid Isolation of Plasmid DNA (Miniscreen) for *Bacillus subtilis*

Materials

SET Buffer: 20% sucrose, 50 mM Tris pH 7.6, 50 mM EDTA.

Lytic Mixture: 1% SDS, 0.2 N NaOH (make fresh mixture every two weeks).

Na(sodium) Acetate Buffer: 3.0 M, pH 4.8.

RNase Buffer: Pancreatic ribonuclease A, 10 mg/ml in 0.1 M Na acetate, 0.3 mM EDTA, pH 4.8, preheated to 80°C for 10 minutes.

Lysozyme: 5 mg/ml in 10 mM Tris HCl pH 7.6, 1 mM EDTA, 10 mM NaCl (TEN Buffer).

Isopropanol.

Ethanol: 70%.

Water (ddH$_2$O):   Sterile double distilled.

Brain Heart Infusion Broth (BHI):   1.5 ml.

Protocol

1. Grow to late log phase of cell growth, a 1.5 ml BHI culture of a plasmid-containing *B. subtilis* strain.

2. Transfer each culture to separate microcentrifuge (microfuge) tubes and spin 1 minute. (All centrifugations will be done in 1.5 ml microfuge tubes.)

3. Resuspend the cell pellet in 1 ml SET Buffer by thoroughly vortexing for 1 minute.

4. Centrifuge the cells for 1 minute and resuspend in 150 μl SET Buffer.

5. Add 20 μl RNase Buffer, 50 μl of lysozyme, and vortex. Add 350 μl of Lytic Mixture at room temperature. Vortex briefly. Solution will become clear.

6. Place on ice bath and incubate 10 minutes.

7. Add 250 μl Na-Acetate Buffer and invert the tube 15 times. Incubate 60 minutes on ice. SDS and denatured chromosomal DNA will precipitate out of solution at this stage.

8. Centrifuge the solution 6 minutes at 4°. Pour the supernatant to a clean microfuge tube (approximately 750 μl).

9. Add an equal volume of isopropanol, invert tubes several times, and centrifuge 10 more minutes at room temperature. Invert tubes and drain to remove isopropanol.

10. Wash the DNA pellet with 1 ml 70% ethanol, and centrifuge 5 minutes at room temperature. Vacuum dry the tubes for 10 minutes. Resuspend in 20 μl of ddH$_2$O.

11. Use 1–2 µl of the "miniscreen" DNA for each restriction digest.

   Note: A second wash with 70% ethanol may be required if restriction digests do not go to completion.

APPENDIX 2
PROTOCOLS

Purification of Phage Lambda DNA

Materials

E. coli AB2500 ($\lambda cI_{857}S_7$).

LB Medium: 1.1 liters.

SM Buffer: 0.02 M Tris-HCl pH 7.5, 0.10 M NaCl, 0.001 M $MgSO_4$, 0.01% gelatin. Autoclave to sterilize and dissolve gelatin.

Chloroform.

DNase I.

Whatman 10 cm Filter Paper.

Celite Slurry: 10 gm Celite in 50 ml of TE Buffer.

TE Buffer: 10 mM Tris-HCl pH 7.6, 1 mM Na_2EDTA.

62.5% CsCl Stock Solution: 25 gm CsCl in 15 ml ddH_2O.

Phenol: Buffer equilibrated.

5M NaCl.

EtOH: 95%.

Ultracentrifuge Rotors and Tubes:
  Beckman Type 35.
  Beckman SW50.1 (or equivalent rotors).

Dialysis Tubing.

Protocol

DAY 1

1. Inoculate 100 ml of LB medium with AB2500 ($\lambda cI_{857}S_7$) and incubate overnight at 30°C with shaking.

DAY 2

1. Inoculate 1 liter of LB medium with the 100 ml overnight culture and incubate at 30°C for 30 minutes with shaking.

2. Shift the growing culture to 43°C for 30 minutes (preferably in a shaking water bath to rapidly change the temperature). This will induce the temperature sensitive lysogen in the strain.

3. Reduce the temperature to 37°C and continue to incubate with shaking for 3 to 5 hours.

4. Harvest the cells by centrifugation (Sorvall HS-4 rotor, 6.5 krpm, 4°C, 15 minutes).
5. Resuspend cells in 100 ml of SM Buffer.
6. Add 10 ml chloroform. Shake slowly at 37°C for 10 minutes.
7. Add 50 μg DNase I. Shake slowly at 37°C for 10 minutes.
8. Let stand overnight at 4°C.

DAY 3

1. Repeat DNase I treatment for 30 minutes, then mix solution well.
2. Centrifuge to separate the phases and remove debris (Sorvall HS-4 rotor, 6.5 krpm, 4°C, 10 minutes). Remove and save the aqueous layer.
3. Pass the aqueous layer through a Celite sandwich in a Buchner funnel:
   10 cm diameter filter.
   Celite slurry.
   10 cm diameter filter.
4. Collect the filtrate. Harvest the phage from the filtrate by centrifugation in the Beckman Type 35 or equivalent rotor (35 krpm, 4°C, 90 minutes).
5. Discard the supernatants and cover the blue-white phage pellets with a total volume of 1.5 ml of TE Buffer. Let stand overnight at 4°C.

DAY 4

1. Pool the resuspended pellets, rinse tubes with 1 ml of TE Buffer, and add this to the pooled phage suspension.
2. Make step gradients of CsCl as indicated below:

	ml 62.5% CsCl	ml ddH$_2$O	density
Solution A	8	4	1.6
Solution B	6	6	1.46
Solution C	4	8	1.32

3. Layer 1 ml each of solutions A, B, and C into 5 ml nitrocellulose ultracentrifuge tubes starting with the least dense solution (C) and underlayering with the denser solutions (B, then A). Gently layer 1–1.5 ml of the phage suspension on top of each gradient.
4. Centrifuge in the SW50.1 (or equivalent), swinging bucket rotor at 35 krpm, 5°C, for 60 minutes. Phage will form a light blue band at the 1.46/1.6 interface (density = 1.505).
5. Puncture the bottom of each tube and collect the phage band.
6. Dialyze overnight at 4°C against 4 liters of TE Buffer.

DAY 5

1. Add 2 ml of buffer-saturated phenol to the phage suspension and roll tube gently 4 hours at room temperature.
2. Centrifuge to separate the phases (Sorvall HS-4 rotor, 6.5 krpm, 4°C, 15 minutes). Remove as much of the lower phenol phase as possible with a pasteur pipet.
3. Repeat the phenol extraction and let stand overnight at 4°C. Remove phenol as in previous step.

DAY 6

1. Dialyse the aqueous phase at 4°C for 6 hours against 4 liters of TE Buffer.
2. Transfer the dialysed DNA solution to a 30 ml Corex tube and add 1/25 volume of 5 M NaCl and 2 volumes of cold 95% EtOH. Mix gently and incubate overnight at −20°C.

DAY 7

1. Collect the DNA precipitate by centrifugation (Sorvall HS-4 rotor, 6.5 krpm, −5°C, 60 minutes). Discard the supernatant, drain the DNA pellet, and dry briefly under vacuum to remove excess EtOH. Cover DNA pellet with 5 ml TEN Buffer.
2. Allow 1 to 2 days for the DNA to resuspend. The solution should become *very* viscous and should be handled with care to prevent shearing of the DNA. The yield should be 7–10 mg DNA/liter culture. Final concentration can be estimated by gel electrophoresis or by determining the A_{260} of a diluted sample. Store at 4°C over a drop of chloroform.

Isolating High Molecular Weight DNA from Yeast

Materials

(Protocol A)

SET Buffer: 20% sucrose, 50 mM Tris-HCl pH 7.6, 50 mM EDTA.

YEPD Medium: 1 liter, (yeast extract, peptone, dextrose).

TEN Buffer: 10 mM Tris-HCl pH 7.6, 1 mM EDTA, 10 mM NaCl.

Dry Ice.

Ethanol: 95%.

Sodium Dodecyl Sulfate (SDS): 25%.

Chloroform/n-amyl Alcohol: 24 to 1 mixture.

RNase Buffer: Pancreatic ribonuclease A, 10 mg/ml, in 0.1 M Na acetate, 0.3 mM EDTA, pH 4.8, preheated to 80°C for 10 minutes.

Pronase: 2 mg/ml in TEN Buffer, preheated to 37°C for 15 minutes.

APPENDIX 2 PROTOCOLS

Additonal materials for Protocol B

Spheroplast Buffer (SB): 0.2 M Tris-HCl pH 9.0, 1 M sorbitol, 0.1 M EDTA, 0.1 M 2-mercaptoethanol.

Zymolyase (Miles Laboratories): 60 units/ml of TEN Buffer.

Phenol/Chloroform: 1 to 1 mixture.

Note: Isolation of yeast DNA can be achieved by disrupting the cell enzymatically or nonenzymatically, using detergents such as SDS. While the latter method is simple and convenient, its effectiveness may vary with different strains of yeast. Therefore, two protocols (A and B) are presented. If lysis cannot be obtained with the freeze-thaw/SDS method, use the enzymatic (spheroplasting) modification presented in Protocol B.

Protocol A

1. Grow 1 liter of yeast cells in YEPD medium to stationary phase of growth (30°C with aeration). Harvest 10 gm (wet weight) of cells in a 250 ml polypropylene centrifuge bottle by spinning the cells at 6.5 krpm (4°C, 5 minutes) in an HS-4 (or GSA) Sorvall rotor.

2. Wash the pellet by resuspending the cells in 60 ml of ddH$_2$O. Centrifuge the cell suspension at 6.5 krpm at 4°C for 5 minutes. Decant the supernatant.

3. Freeze the cell pellets by placing in the dry ice-ethanol bath for about 5 minutes. Thaw the pellets by placing the bottles in warm water. Repeat this step 2 more times.

4. Resuspend the thawed pellets in 20 ml of SET Buffer and 1/25 volume of 5 M NaCl.

5. Add 1/10 volume of 25% SDS solution, mix, and let stand at room temperature for 1 hour. Heat the lysis mixture to 65°C for 1 hour.

6. Cool the lysate to room temperature and add an equal volume of chloroform-n-amyl alcohol. Gently mix until the mixture is homogeneous.

7. Separate the two phases by centrifugation at 6.5 krpm, 4°C for 5 minutes (HS-4 rotor). Using a wide-bore pipet, transfer the upper (aqueous) phase to a clean centrifuge bottle and repeat the chloroform extraction (Steps 6 and 7).

8. Precipitate the DNA by transferring the upper phase to a centrifuge bottle containing 2 volumes of ice-cold 95% ethanol. Mix gently and spool out the chromosomal DNA on a glass rod or bent pasteur pipet.

9. Resuspend the DNA in 2 ml of TEN Buffer (in a 30 ml Corex tube) and add 0.5 ml of RNase Buffer. Incubate at 37°C for 1 hour. Next add 0.5 ml of pronase and continue the incubation another 2 hours.

10. Deproteinize the solution by adding an equal volume of chloroform-n-amyl alcohol and repeat the extraction as described in Steps 6 and 7.

11. Transfer the upper phase to a clean 30 ml Corex tube. Add 1/25 volume of 5 M NaCl and 2 volumes ice-cold 95% ethanol. Mix the solution gently and chill on ice for 5 minutes. Spool out the DNA on a glass rod or bent pasteur pipet.
12. Resuspend the DNA in 1 to 5 ml of TEN Buffer and store over 0.1 ml of chloroform at 4°C.

Protocol B

1. Grow, harvest, and wash yeast cells as described in Steps 1 and 2 of Protocol A.
2. Resuspend cell pellet in Spheroplast Buffer at a density of 0.25 gm of cells/ml of buffer (approximately 40 ml). Add 2 ml of Zymolyase to the suspension and incubate with occasional mixing for 1 hour at 30°C. (Lyticase can also be used in place of Zymolyase at this step.)
3. Harvest the spheroplasts by centrifugation at 4.5 krpm for 5 minutes, 4°C (4000 g), and gently resuspend the pellet in 40 ml of Spheroplast Buffer. To ensure complete lysis and good DNA yield, make an even cell suspension by breaking up all clumps of cells.
4. Add an equal volume of TEN Buffer, 2 ml of SDS (to give a final concentration of 1%), and 0.5 ml of RNase Buffer. Mix gently and incubate at 37°C for 2 hours. Add 1 ml (2 mg) of pronase and continue the incubation for another 2 hours. Gently swirl the mixture periodically.
5. Heat the mixture to 65°C for 30 minutes and cool to room temperature.
6. To extract residual proteins, add an equal volume of phenol/chloroform and mix gently until the two solutions are homogeneous (about 15 minutes).
7. Separate the two phases by centrifugation (6.5 krpm for 15 minutes, 4°C in an HS-4 rotor).
8. Using a wide-bore pipet, transfer the aqueous phase to a clean 250 ml centrifugation bottle and repeat Steps 6 and 7 using chloroform/n-amyl alcohol.
9. Add 1/25 volume 5 M NaCl and divide the aqueous phase equally between 2 clean centrifuge bottles. Precipitate the DNA with 2 volumes of cold 95% EtOH. Let the solution sit on ice for 5 minutes and swirl occasionally. Spool out the DNA on a glass rod and dissolve in 50 ml of TEN Buffer.
10. Add 0.1 ml of RNase Buffer and incubate at 37°C for 1 hour.
11. Extract the RNase and any residual protein by repeating Steps 6 and 7. Add 1/25 volume of 5M NaCl and precipitate DNA with 2 volumes of cold 95% EtOH. Chill on ice for 5 minutes with occasional stirring.
12. Spool the DNA on a glass rod and dissolve in 1 to 5 ml of TEN Buffer. Store at 4°C over 0.1 ml of chloroform. Determine the A_{260}/A_{280} ratio for an estimate of concentration and purity.

APPENDIX 2
PROTOCOLS

Isolation of Yeast Plasmid DNA*

Materials

Yeast Minimal Medium: 40 ml yeast culture grown to 5×10^7 cells/ml.

Spheroplast Buffer: 0.2 M Tris-HCl pH 9.0, 1.2 M sorbitol, 0.1 M EDTA, 0.1 M 2-mercaptoethanol.

SCE Buffer: 1.0 M sorbitol, 0.1 M sodium citrate pH 5.8, 60 mM EDTA.

Zymolyase 60,000: (Miles Laboratories.)

ST Buffer 25% sucrose, .05 M Tris-HCl, pH 8.0.

Lysis Buffer: 50 mM Tris-HCl, 20 mM EDTA, 1% (w/v) SDS. Adjust to pH 12.45 at 23°C.

2 M Tris-HCl pH 7.0.

pHydrion Paper Strips.

Solid NaCl.

Phenol: Saturated with 3% (w/v) NaCl.

TE Buffer: 10 mM Tris-HCl pH 7.6, 1 mM EDTA.

Ethanol: 95%.

Ethanol: 70%.

Protocol

1. Harvest yeast cells in a 50 ml polypropylene tube by centrifugation for 5 minutes in a tabletop centrifuge.

2. Wash cell pellet once by resuspending in 40 ml ddH$_2$O. Pellet by centrifugation.

3. Resuspend cell pellet in 20 ml Pretreatment Buffer. Incubate at room temperature 10 minutes. Pellet by centrifugation.

4. Wash cell pellet twice by resuspension in 25 ml of SCE Buffer.

5. Resuspend cell pellet in 25 ml SCE Buffer (cell density should not exceed 10^8 cells/ml) and add Zymolyase 60,000 to a final concentration of 100 µg/ml.

6. Incubate at 37°C for 25 minutes.

7. Pellet spheroplasts by centrifugation in a tabletop centrifuge for 5 minutes.

8. Gently resuspend the spheroplast pellet in 0.5 ml ST Buffer.

*From DeVenish and Newlon (1982). Gene 18, 277.

9. Add 9.5 ml Lysis Buffer in dropwise fashion while stirring at 100 rev/min with a Teflon-coated magnetic stir bar. Stir for 90 seconds.
10. Incubate lysate 25 minutes at 37°C.
11. Adjust pH to 8.5–8.9 (as measured with pH paper) by the addition of 2 M Tris-HCl pH 7.0, while stirring. Stir for 2 minutes.
12. Measure volume and add solid NaCl to give a final concentration of 3% (w/v). Incubate at room temperature 30 minutes.
13. Add an equal volume of NaCl-saturated phenol. Mix by stirring 10 seconds at 300 rpm, then for 2 minutes at 100 rpm.
14. Separate the phases by centrifugation. Recover the aqueous phase and add 2 volumes of cold 95% EtOH to precipitate nucleic acids.
15. Incubate 2 hours at −20°C.
16. Collect the precipitated material by centrifugation (Sorvall HS-4 rotor, 6.5 krpm, −5°C, 30 minutes).
17. Wash the pellet once in 10 ml of cold 70% EtOH to remove excess salt. Following centrifugation (6.5 krpm, −5°C, 10 minutes), pour off the ethanol wash, dry the pellet under vacuum (10 to 30 minutes), and resuspend in 50 μl TE Buffer. Store at 4°C.

Rapid Method for Isolating Plasmid DNA from Yeast

Plasmid DNA isolated by this procedure can be used to transform *E. coli*; however, the DNA is not concentrated or pure enough to perform restriction enzyme analysis.

Materials

YM Medium: 5 ml of yeast minimal medium supplemented with amino acids and vitamins as required.

Spheroplast Buffer (SB):   0.2 M Tris-HCl pH 9.0, 1 M sorbitol, 0.1 M EDTA, 0.1 M 2-mercaptoethanol.

Zymolase 60,000:   1 mg/ml in TEN Buffer (Miles Laboratories).

SET Buffer: 20% sucrose, 50 mM Tris-HCl pH 7.6, 50 mM EDTA.

Lytic Mixture: 1% SDS, 0.2 N NaOH (make fresh every two weeks).

Na (sodium) Acetate Buffer: 3.0 M, pH 4.8.

RNase Buffer: Pancreatic ribonuclease A, 10 mg/ml in 0.1 M Na acetate, 0.3 mM EDTA, pH 4.8, preheated to 80°C for 10 minutes.

Isopropanol.

Ethanol: 70%.

Water (ddH$_2$O):   Sterile double distilled.

Protocol

1. Inoculate 5 ml of YM with a plasmid-containing strain of yeast. Grow the culture to log phase (5×10^7 cells/ml) by incubating at 30°C with aeration, for about 36 hours.
2. Check the A_{600} value for the culture and adjust the cell density to give an A_{600} of 1.5.
3. Harvest 5 ml of yeast cells by centrifugation in a tabletop centrifuge.
4. Wash the cells twice with distilled water and pellet the cells as in Step 3.
5. Resuspend the cell pellet in 5 ml of SB and add $5\mu l$ of Zymolyase. Incubate for 1 hour at 30°C with gentle agitation.
6. Collect cells as in Step 3 and resuspend the cell pellet in $150\mu l$ of SET Buffer.
7. Add $5\mu l$ of RNase Buffer, $350 \mu l$ of Lytic Mixture (at room temperature), and vortex briefly. Place on ice for 10 minutes.
8. Add $150 \mu l$ Na Acetate Buffer and invert the tube several times. Incubate 30 minutes on ice.
9. Centrifuge the solution 5 minutes at 4°C in a microcentrifuge. Pipet the supernatant to a clean microfuge tube (approximately $700 \mu l$).
10. Add an equal volume of isopropanol, invert tubes several times and centrifuge 5 more minutes at room temperature. Invert tubes and drain to remove isopropanol.
11. Wash the DNA pellet with 1 ml 70% ethanol, and centrifuge 5 minutes at room temperature. Vacuum dry the tubes for 10 minutes. Resuspend in $20\mu l$ of ddH$_2$O.
12. Use 5 to 10 μl of the "miniscreen" DNA for each *E. coli* transformation.

Drosophila DNA Isolation

Materials

Lysis Buffer (LB): 0.15 M NaCl, 0.015 M Na Acetate, 0.1 M EDTA, pH 8.0.

Homogenization Buffer (HB): 0.01 M NaCl, 0.2 M sucrose, 0.01 M EDTA, 0.03 M Tris-HCl, pH 8.0.

TEN Buffer: 10 mM Tris-HCl pH 7.6, 10 mM NaCl, 1 mM EDTA.

Proteinase K (Beckman).

SDS (Sodium Dodecyl Sulfate): 25%.

5 M Sodium Perchlorate.

Chloroform/n-amyl alcohol: 24 to 1.

RNase Buffer: Pancreatic ribonuclease A, 10 mg/ml in 0.1 M Na acetate, 0.3 M EDTA pH 4.8, preheated to 80°C for 10 minutes.

Ethanol: 95%.

Protocol

1. Grind 5 gm frozen (−70°C) flies to a fine powder using mortar and pestle that are kept cold in dry ice.
2. Suspend powdered fly bodies in 20 ml Homogenization Buffer (HB) in large dounce and dounce 20 strokes with an "A" (small) pestle. Repeat with 2 to 3 strokes using a "B" (large) pestle.
3. Filter twice through miracloth to remove large particulate matter. Spin filtrate at 7 krpm (4°C) for 10 minutes in HB-4 rotor to pellet nuclei. Pour off and discard supernatant. Resuspend nuclei in 5 ml HB and repeat spin. Pour off supernatant and repeat wash/spin once or twice more until supernatant is clear.
4. Suspend nuclei in 4.5 ml Lysis Buffer and transfer to small dounce. Dounce 20 strokes. Lyse nuclei by adding 0.5 ml 25% SDS and mix very gently. Add 3 mg Proteinase K and incubate at 37°C for 1 hour.
5. Extract protein. Add 1.5 ml 5 M sodium perchlorate and add chloroform/n-amyl alcohol to give volume equal to that of aqueous phase. Mix gently 10 minutes. Solution forms emulsion. Separate phases by centrifugation (10 minutes, 4°C, 10 krpm, Sorvall HB-4 rotor). Extract aqueous phase 3 times additionally as above. Back extract each organic phase using 5 ml LB + 1 ml Na perchlorate. Precipitate nucleic acids by adding two volumes cold EtOH and spool out onto glass rod. Resuspend in 4 ml TEN Buffer (sample will now contain about equal amounts of DNA and RNA). Due to eye pigment that copurifies with the DNA, the solution will be pink. This pigment does not affect the physical properties of the DNA or inhibit restriction enzyme cleavage.
6. Add 20 μl of RNase Buffer and incubate at 37°C for 1 hour. Add 0.2 ml 5 M NaCl and extract protein using an equal volume of chloroform/n-amyl alcohol. Gently mix for 10 minutes. Spin to separate phases at 10 krpm for 10 minutes. Remove aqueous phase and extract another 2 times or until the interface between the two phases remains clear. Precipitate DNA with two volumes cold EtOH and spool out onto glass rod. Resuspend in 2 ml TEN Buffer. Usual yield is about 500–600 μg DNA with no detectable RNA. $A_{260}/A_{280} = 1.8$ to 2.0.

APPENDIX 2
PROTOCOLS

Extraction of Restriction Fragments from Low-melt Agarose Gels*

Useful for DNA fragments ranging from 100 bp to 4 kb.

Materials

Low-melt Agarose (Sigma).
Hexadecyltrimethylammonium Bromide (Hab) (Sigma H-5882).
Antifoam A Emulsion (Sigma A-5758).

---
*J. Langridge, P. Langridge, P. L. Bergquist (1980). Anal. Biochem. 103, 264–271.

APPENDIX 2
PROTOCOLS

n-Butanol (Sigma BT-105).

0.2 M NaCl.

Chloroform.

Phenol: Buffer equilibrated.

ddH$_2$O: Sterile.

Dowex Resin: AG 50W-X8. (See Appendix 1.)

Ethanol: 95%.

Ethanol: 70%.

Ethidium Bromide: 4 μg/ml, made up with water.

Protocol

Stock solutions

1. H$_2$O saturated butanol/Hab.
2. Butanol saturated H$_2$O/antifoam.
    a. Mix 60 ml H$_2$O together with 50 ml butanol. Let phases separate for 5 minutes at room temperature.
    b. Transfer 10 ml butanol-saturated H$_2$O (bottom phase) and 10 ml H$_2$O-saturated butanol (top phase) into labeled tubes.
    c. Add 25 μl antifoam A to tube containing butanol-saturated H$_2$O, and dissolve 0.5 gm Hab in tube containing H$_2$O-saturated butanol.
    d. Add solutions prepared in Step "c" back to biphasic solution prepared in Step "a," mix thoroughly, and store at room temperature as the biphasic solution.

Electrophoresis

Prepare either horizontal or vertical gels. Minigels are suitable for extractions involving less than about 20 μg DNA. Run the gel at 4°C; do not allow the buffer to heat up or the gel will melt. Extractions of small agar slices can be performed in microfuge tubes; larger extractions should be performed in silanized test tubes.

Band Extraction

1. To cut out the desired band, stain the gel from 5 to 15 seconds in EtBr and use a UV transilluminator to visualize the DNA.
2. Melt gel slice at 65°C. Measure gel volume.
3. Add equal volumes of aqueous phase and butanol phase of biphasic extraction reagent.
4. Vortex tube 10 seconds and spin in a tabletop centrifuge 5 krpm for 3 minutes at room temperature.
5. Transfer butanol (upper phase) to a separate tube.

6. Repeat Steps 3–5 two more times. Pool the three butanol extracts.
7. To pooled butanol add ¼ volume 0.2 M NaCl, vortex, and centrifuge at room temperature for 5 minutes.
8. Transfer the lower aqueous phase to a 15 ml Corex tube. (All subsequent centrifugations will be performed using 15 ml Corex tubes and the HS-4 or SS34 Sorvall rotors.)
9. Repeat Steps 7 and 8 once more and pool the two aqueous extracts.
10. Add an equal volume of buffer-equilibrated phenol to aqueous phase; vortex briefly. Add the same volume of chloroform, vortex 30 seconds, then centrifuge at 4°C for 5 minutes. Remove and save the aqueous (upper) phase.
11. Add an equal volume chloroform; vortex briefly. Centrifuge 5 krpm at 4°C for 5 minutes. Remove and save the aqueous (upper) phase.
12. Remove residual EtBr by the addition of 1/5 volume of Dowex AG 50W-X8 resin. Mix 3 times over 5 to 10 minutes. Remove and save the supernatant.
13. Add two volumes of −20°C 95% EtOH and precipitate 1 hour at −20°C.
14. Collect the DNA precipitate by centrifugation (−5°C, 30 minutes, 6.5 krpm).
15. Wash the DNA pellet by addition of 1 ml cold 70% EtOH. Invert tube several times, then centrifuge 5 minutes.
16. Remove the EtOH supernatant from the DNA pellet. Dry the pellet and resuspend in an appropriate volume of ddH$_2$O or TEN (generally 20–200 µl).

Preparative Agarose Gel Electrophoresis

Materials

Materials for agarose gel electrophoresis.

DNA Extraction Buffer:

ammonium acetate	3.86 gm	(0.5 M)
magnesium acetate	0.214 gm	(10 mM)
Na$_2$EDTA	3.72 gm	(0.1 mM)
SDS (sodium dodecyl sulfate)	0.1 gm	(0.1%)
ddH$_2$O to	100 ml volume.	

Note: Ammonium acetate is unstable. Refrigerate the dry chemical and use freshly made solutions.

Ethanol: 95%.

2 M NaCl.

Dialysis Tubing.

APPENDIX 2
PROTOCOLS

Protocol

1. Prepare an agarose gel having a single well. This may be achieved by using a specially constructed comb, or by inverting the normally-used comb such that the teeth point up instead of down. The following factors are adjusted to suit the investigators' needs:

    a. *Percent agarose*—use from 0.8%–1.4% agarose, depending on the sizes of DNA fragments being isolated, the degree of DNA band separation required, and the amount of DNA applied to the gel. A high percent agarose is generally chosen in order to minimize trailing of the DNA bands, especially when large amounts of DNA are applied to the gel. However, large DNA fragments are easier to extract from lower percent agarose gels. Typical conditions are a 1.0% agarose gel for isolating a 2.5 kb fragment and a 1.4% agarose gel for a 500 bp fragment.

    b. *Gel size*

       1) length—the longer the distance traveled, the greater the separation of DNA fragments. A glass plate of 6" in length is routinely used for most DNA fragment isolations.

       2) width—approximately 4–5". The wider the well, the greater the resolution of individual fragments.

       3) thickness—for 50 μg of DNA or less, use a 1.5 mm spacer; for 100 μg of DNA or more, use a 4 mm spacer.

2. Apply DNA in dye-glycerol solution (reaction stop mix) to gel. Layer the DNA at the base of the well as carefully and evenly as possible. The less mixing of the DNA-dye with the running buffer, the better the resolution within the gel.

3. Running time—a function of the size of the DNA fragment to be isolated, percent agarose, gel length, gel thickness, and applied voltage. Typical conditions are as follows:

    a. 2.5 kb insert; 1.1% agarose; 6" × 1.5 mm; 150 volts (30–40 mAmp) = 3 hours running time.

    b. 500 bp insert; 1.4% agarose; 6" × 4 mm; 80 volts (40–50 mAmp) = 4 hours running time.

4. Localizing the DNA fragments:

    a. Demarcate the top of the gel by making a diagonal incision with a razor blade.

    b. Cut off the sides of the gel, taking a narrow region from each side that contains DNA.

    c. Stain the excised sides of the gel in a 4 mg/ml solution of ethidium bromide (in water) for 1–2 minutes. Destain briefly in water.

d. Visualize the DNA with a UV light and make identations in the gel where the DNA is located. Example:

APPENDIX 2
PROTOCOLS

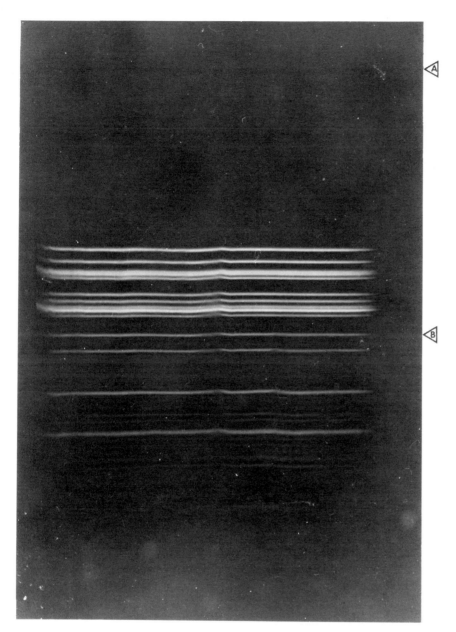

(The figure shown above is a *Hae*III digestion of the *Pst*I recombinant of pBR322, which carries the araB and araC genes of the arabinose operon. The gel is 7.5% acrylamide, 0.4 mm thick and 6 inches long. Triangle A indicates the top of the gel while triangle B indicates the *Hae*III fragment containing the promoters for araB and C.)

e. Realign the marked agarose strips with the unstained gel. Excise the region of the gel containing the DNA fragment of interest.

5. Place the agarose strip on a clean glass plate and mince as fine as possible with a razor blade. Moisten the agarose occasionally with DNA Extraction Buffer.
6. Place minced agarose into a 4 ml polyallomer centrifuge tube (for SW50.1 rotor). More than one tube may be required.
7. Fill tube(s) with DNA Extraction Buffer (2–3 ml).
8. Suspend the agarose fragments in the Extraction Buffer with a spatula or sterile glass rod. The size of the fragments may be further reduced by crushing them against the side of the tube.
9. Seal the top with Saran wrap.
10. Incubate at 65°C for 8 to 12 hours. During this time the DNA will diffuse out of the agarose and into the Extraction Buffer.
11. Collect the DNA by centrifuging the tubes in the Beckman SW50.1 rotor for 30 minutes, 30 krpm, 20°C. Pour off and save the aqueous DNA solution.
12. Resuspend the agarose pellet in 3.5 ml of DNA Extraction Buffer.
13. Incubate at 65°C for 1 to 2 hours.
14. Repeat Step 11.
15. Pool the DNA solutions.
16. Add 2 volumes of cold 95% EtOH. Place at −20°C for at least 1 hour.
17. Centrifuge to collect the DNA precipitate (Sorvall HS-4 rotor, 6.5 krpm, −5°C, 1 hour. Discard supernatant.
18. Resuspend DNA pellet in 0.5–1.0 ml sterile ddH$_2$O.
19. Dialyze against ddH$_2$O 12 to 24 hours at room temperature to remove salt and SDS.
20. Transfer DNA solution to a microfuge tube. Centrifuge 2 minutes. Carefully remove the supernatant, avoiding any agarose pellet in the tube. Rinse the tube once by the addition of 0.2 ml of ddH$_2$O. Centrifuge, and again carefully withdraw the supernatant. Pool the DNA solutions.
21. Reduce volume to 100–200 μl. This can be done by evaporation or by ethanol precipitation. To precipitate, add 1/10 volume of 2 M NaCl and 2 volumes of cold 95% EtOH, then repeat Step 17.

Electroelution of DNA Fragments from Polyacrylamide Gels*

Materials

Dialysis Tubing.

Electroelution Box.

TBE (0.1X): Tris-Borate buffer diluted 100-fold from 10X stock.

2.5 M Ammonium Acetate.

Ethanol: 95%.

Ethanol: 70%.

TEN Buffer.

---

*H. O. Smith (1980). Methods in Enzymology 65, Academic Press, New York, pp. 371–373.

Protocol

1. Cut a piece of dialysis tubing slightly longer than the gel slice and fill with 0.1X TBE buffer.
2. Place gel slice in bag, seal the end, and place the bag into the electroelution box parallel to the electrodes.
3. Add 0.1X TBE Buffer until the bag is almost immersed.
4. Apply 150 V (approximately 45 mA) for 30 minutes.
5. Reverse polarity by switching the electrodes and back-electrophorese for 5 seconds to remove DNA bound to the sides of the dialysis tubing.
6. Carefully remove DNA from the bag and place into a 30 ml silane-treated Corex tube.
7. Add 1/10 volume of 2.5 M ammonium acetate and 2 volumes of EtOH.
8. Collect the DNA precipitate by centrifugation ($-5°C$, 30 minutes, 6.5 krpm) using a Sorvall HS-4 rotor.
9. Wash DNA pellet once in 70% EtOH, dry the pellet, and resuspend in appropriate volume (20–200 μl) of TEN Buffer.

Transformation of *E. coli* χ1776*

Materials

LB + DAPT: Luria broth supplemented with 100 μg/ml diaminopemelic acid (DAP), 40 μg/ml thymidine (T).

TN Buffer: 10 mM Tris-HCl pH 8.0, 0.8% NaCl.

TNC Buffer: 75 mM $CaCl_2$ made up in TN Buffer.

100 mM NaCl.

Note: Although this EK2 host can be transformed by the standard $CaCl_2$ transformation protocols, the strain is prone to lysis and reduced transformation efficiencies will be observed. The procedure below has been designed to optimize transformation frequency of this strain.

Protocol

1. Dilute 2 ml of an overnight culture of χ1776 grown in LB + DAPT at 37°C into 18 ml fresh LB + DAPT. Incubate 3.5 hours at 37°C with shaking.
2. Harvest the cells by centrifuging the culture at room temperature in a tabletop centrifuge (3 krpm for 5 minutes).
3. *Gently* resuspend the cell pellet in 10 ml 100 mM NaCl and centrifuge again at room temperature.

---

*M. Inone and R. Curtiss, in "Molecular Cloning of Recombinant DNA", W. A. Scott and R. Werner, eds. Academic Press, New York, 1977, p. 248.

APPENDIX 2
PROTOCOLS

4. Gently resuspend the cell pellet in 10 ml of TNC Buffer and incubate at room temperature for 20 minutes. Harvest the cells as described in Step 2.

5. Gently resuspend the cell pellet in 2 ml of TNC Buffer. Chill on ice in a glass tube for 2 minutes.

6. Mix 0.2 ml of cells with 0.1 ml of DNA in TN Buffer. Incubate on ice for 20 minutes.

7. Heat to 42°C for 1 minute, then place on ice for 10 minutes.

8. Plate directly on selective plates.

M13 Transformation

Materials

Competent JM101 Cells: Bacterial host for bacteriophage M13.

YT media:
  Agar plates.
  Soft agar overlay.

100 mM IPTG: 23.8 mg isopropyl-β-D-thiogalactopyranoside · ½ dioxane (in 1 ml sterile ddH$_2$O).

2% X-gal: 25 mg 5-bromo-4-chloro-3-indolyl-β-D-galactopyranoside (in 1.25 dimethylformamide).

M13 DNA Sample.

Protocol

1. Add 1–10 μl of M13 DNA sample to 200 μl of competent JM101 cells. Incubate on ice from 5 to 60 minutes.

2. Melt YT soft agar and aliquot 3 ml per transformation into sterile glass test tubes. Cool to 43–45°C.

3. To the tubes containing the cooled YT soft agar, add 10 μl of 100 mM IPTG plus 50 μl 2% X-gal. Vortex gently to mix ingredients.

4. Add the competent cells plus DNA directly to the soft agar. Mix gently and incubate at 43°C–45°C for 2 minutes.

5. Pour the soft agar overlay (containing the transformed cells) onto the surface of a YT agar plate. Rotate the plate to evenly distribute the liquid overlay, then allow the plate to stand 5 to 15 minutes to allow the soft agar to harden.

6. Incubate at 37°C. The nontransformed competent cells will form an indicator lawn, and transformed cells will give rise to plaques visible in 4 to 5 hours. Blue color will become obvious after 5 to 7 hours and will progressively darken.

7. Recombinant (white) plaques can be picked with a sterile loop and used to inoculate 2 ml cultures of YT medium containing about 10^7 cells/ml JM101 (or similar strain sensitive to M13).

8. After 5 to 7 hours growth at 37°C, the infected cells can be harvested and the double-stranded RF DNA can be isolated and analyzed by the miniscreen procedure described in Exercise 5. Note that the supernatant will contain the packaged single-stranded phage and should be saved if DNA sequencing of the recombinant is desired.

APPENDIX 2
PROTOCOLS

Bacteriophage Lambda *In Vitro* Packaging

Materials

E. Coli BHB2688 N205 recA (λ imm$_{434}$cI$_{ts}$ b2 red3 Eam4 Sam7)/λ.

E. coli BHB2690 N205 recA (λ imm$_{434}$cI$_{ts}$ b2 red3 Dam15 Sam7)/λ.

(Both strains should be temperature sensitive with no growth at 42°C.)

Complementation Buffer: 40 mM Tris-HCl pH 8.0, 10 mM spermidine-HCl, 10 mM putrescine-HCl, 0.1% 2-mercaptoethanol, 7% dimethyl sulfoxide.

Liquid Nitrogen.

10 mM ATP.

DNase I: 1 mg/ml.

SMC Buffer:

Na$_2$HPO$_4$	0.7 gm
KH$_2$PO$_4$	0.3 gm
NaCl	0.05 gm
ddH$_2$O to	100 ml volume.
Autoclave, cool, then add	
1 M MgCl$_2$	0.1 ml
0.1 M CaCl$_2$	0.1 ml.

Chloroform.

LB Medium.

LB Medium + 0.4% Maltose.

For transduction of phage:
  0.01 M MgSO$_4$

LAM Agar Plates:
  LB agar plates containing 2.5 gm MgSO$_4 \cdot$7H$_2$O per liter.

LAM Soft Agar:
  Same as LAM plates but containing only 0.6% agar.

APPENDIX 2
PROTOCOLS

Protocol

I. *To make a packaging mix**

DAY 1

1. Grow a single colony of each strain overnight at 30°C in 4 ml of LB medium.

DAY 2

1. Inoculate two 100 ml cultures of LB medium with 2 ml of the stationary phase overnight cultures.
2. Grow cultures at 30°C with good aeration to $A_{600} = 0.3$ (1.5×10^8 cells/ml).
3. Shift cultures to a 45°C water bath for 15 minutes without agitation.
4. Shift cultures to 37°C and incubate with vigorous aeration for 3 hours.
5. Remove a 1 ml sample from each culture and add a few drops of chloroform. Cultures should lyse. If not, the phage have not been sufficiently induced.
6. If both cultures lysed when tested in Step 5, combine the two cultures in a 250 ml plastic centrifuge bottle and centrifuge (Sorvall GSA rotor, 5 krpm, 7°C, 8 minutes).
7. Resuspend the cell pellet in 10 ml of LB medium, transfer to a 50 ml polypropylene tube, and centrifuge (Sorvall SS34 rotor, 3 krpm, 7°C, 5 minutes).
8. Pour off and discard the supernatant. Remove all excess liquid with a pasteur pipet. Resuspend the cell pellet in 0.4 ml of Complementation Buffer.
9. Estimate total volume of cells and buffer and adjust to 1 mM ATP.
10. Dispense 20 µl aliquots into microfuge tubes. Freeze in liquid nitrogen (*not* dry ice-ethanol). The prepared packaging mix can be stored at −70°C for at least 1 year.

II. *In vitro packaging*†

1. Remove packaging mix from −70°C storage and thaw on ice.
2. Add 2 µl (~600 ng) of ligated DNA sample (can be in Ligation Buffer) and mix gently but thoroughly.
3. Incubate at 37°C for 30 minutes.
4. Add 0.5 ml of SMC Buffer.
5. Add 5 µl of 1 mg/ml DNase I and mix well.
6. Incubate at 37°C for 40 minutes. Viscosity of the solution should decrease.
7. Add a drop of chloroform, mix gently, and centrifuge for 30 seconds.

---

*Collins (1980). In Methods in Enzymology 68, R. Wu, ed., Academic Press, New York, p. 309.
†Ibid.

8. The DNA sample is now packaged and can be saved over a drop of chloroform at 4°C and treated in the manner of a phage lysate.

III. *Transduction of cosmids.* *

1. Grow *E. coli* HB101 or a comparable recipient strain to late log phase in LB + 0.4% maltose at 30°C with shaking. The maltose induces the synthesis of the phage receptor proteins on the surface of the bacteria.
2. Add 0.4 ml of the packaged DNA to 1 ml of cells.
3. Adsorb 15 minutes at room temperature.
4. Plate 200 μl, 50 μl, and 10 μl directly onto LB plates containing the appropriate antibiotic to select for cells containing the cosmid. Incubate at 30°C.
5. Add 1 ml of LB medium to the remaining cells and incubate for 2 hours at room temperature.
6. Plate 200 μl, 50 μl, 10μl, and 2 μl of the recovered cells onto LB plates containing the appropriate antibiotic. Incubate at 30°C.

IV. *Transduction of lambda vectors*†

1. Grow *E. coli* HB101 or comparable indicator strain to late log phase in LB + 0.4% maltose.
2. Harvest by centrifugation and resuspend in the same volume of 0.01 M $MgSO_4$.
3. Add 0.4 ml of the packaged DNA to 1 ml of the cell suspension.
4. Adsorb 15 minutes at room temperature.
5. Typical packaging efficiencies for λ vectors range from 10^4–10^8 plaque-forming units per μg vector DNA. Appropriate dilutions of the transduced cell sample should be mixed with 3 ml of liquid LAM soft agar, poured onto an LAM agar plate, and allowed to solidify.
6. Incubate the plates at 37°C to allow the growth of the bacterial lawn and the formation of phage plaques.

Agrobacterium tumefaciens Transformation‡

Materials

YEB Medium:

Nutrient broth	13.3 gm
Yeast extract	1.0 gm
Sucrose	5.0 gm
Anhydrous $MgSO_4$	.24 gm
ddH_2O to	1 liter

---

*Ibid.
†Hohn (1980). In Methods in Enzymology 68, R. Wu, ed., Academic Press, New York, p. 299.
‡Procedure courtesy of Dr. C. I. Kado, Department of Plant Pathology, University of California, Davis.

**APPENDIX 2
PROTOCOLS**

523 Medium:

Sucrose	10.0 gm
Casein hydrolysate	8.0 gm
Yeast extract	4.0 gm
K_2HPO_4	2.0 gm
$MgSO_4$	0.15 gm
ddH_2O to	1 liter
pH to 7.1	

Dry Ice-Ethanol Bath.

Plasmid DNA.

Protocol*

1. Grow recipient *A. tumefaciens* strain in YEB overnight at 30°C and inoculate into 5 ml YEB at 10^7 cells/ml. Grow at 30°C with shaking until a cell concentration of 4×10^8 cells/ml is reached.

2. Harvest cells by centrifugation at 5 krpm, 4°C. Resuspend cell pellet in 1 ml YEB.

3. Concentrate cells by centrifugation (5 krpm, 4°C) and decant the supernatant.

4. Resuspend the cell pellet in the remaining liquid. Store on ice until the pellet can be frozen in a dry ice-ethanol bath (takes approximately 2 minutes).

5. Thaw at 37°C for 3 minutes and aliquot 0.2 ml of cells into 1.5 ml sterile tubes.

6. Add 90 µl YEB + 10 µl DNA (purified plasmid or miniscreen DNA) for each transformation.

7. Place tubes in dry ice-ethanol for 2 minutes and then thaw at 37°C for 25 minutes.

8. Incubate at 30°C with no shaking for 1 hour.

9. Add transformed cells to 5 ml of 523 medium and incubate at 30°C with shaking for 1 hour.

10. Harvest the culture by centrifugation as described in Step 2 and pour off supernatant. Resuspend the cells in 200 µl of 523 medium.

11. Plate 100 µl on appropriate selective plates. Incubate 30°C for 48 hours.

Transformation of *Bacillus subtilis* with Plasmid DNA*

Materials

All solutions should be autoclaved 20 minutes, then stored at room temperature.

---

*Protocol provided courtesy of Dr. R. H. Doi, Department of Biochemistry, University of California, Davis.

APPENDIX 2

PROTOCOLS

SPI Salts:*
- $(NH_4)_2SO_4$ 1 gm
- K_2HPO_4 7 gm
- KH_2PO_4 3 gm
- NaCitrate·2 H_2O 0.5 gm
- $MgSO_4·7 H_2O$ 0.1 gm
- ddH_2O to   500 ml volume

$CaCl_2$:
- $CaCl_2$ 0.28 gm
- ddH_2O to   500 ml volume

$MgCl_2$:
- $MgCl_2$ 2.54 gm
- ddH_2O to   50 ml volume

Glucose:
- Glucose   25 gm
- hot ddH_2O to   50 ml volume

CAYE:
- Casamino acids   1.0 gm
- Yeast extract   5.0 gm
- ddH_2O to   50 ml volume

EGTA:†
- Ethyleneglycol-bis-(β-aminoethyl ether)—N,N'-tetraacetic Acid   1.9 gm
- ddH_2O to   50 ml volume

SPI Medium:‡
- SPI Salts   100 ml
- Glucose   1 ml
- CAYE   1 ml

SPII Medium:§
- SPI Salts   100 ml
- Glucose   1 ml
- CAYE   1 ml
- $CaCl_2$ 1 ml
- $MgCl_2$ 1 ml

Protocol

DAY 1

1. Inoculate a 5 ml culture of SPI medium with a single *B. subtilus* colony. Incubate overnight at 37°C with rapid shaking.

---

*Dubnau and Davidoff-Abelson (1971). J. Mol. Biol. 56, 209.
†Gryczan et al. (1978). J. Bacteriol. 134, 318.
‡Dubnau and Davidoff-Abelson (1971). J. Mol. Biol. 56, 209.
§Ibid.

APPENDIX 2

PROTOCOLS

DAY 2

1. Inoculate 50 ml of SPI medium with 0.5 ml of the overnight culture. Incubate at 37°C with rapid shaking.
2. Monitor growth by absorbance until the culture enters stationary phase.
3. Add 5 ml of this early stationary phase culture to 45 ml prewarmed SPII medium. Shake slowly at 37°C for 90 minutes.
4. Add 0.5 ml EGTA and continue 37°C incubation for 5 to 10 minutes.
5. Add 0.1–0.5 ml of cell suspension to 1–3 μg of DNA in a small sterile glass tube. Shake at 37°C.
6. After 90 minutes of incubation, plate 0.05–0.10 ml onto dry agar plates containing appropriate selective medium. Incubate at 37°C.
7. To freeze cells for later use, do not add EGTA after Step 4. Cool on ice and harvest by centrifugation (Sorvall HS-4 rotor, 6.5 krpm, 4°C, 10 minutes).
8. Gently resuspend the cells in SPII containing 10% glycerol.
9. Freeze the cell suspension rapidly in dry ice-ethanol. Store at -70°C.
10. To prepare for transformation, thaw the cells rapidly at 42°C. Add 4 ml SPII per ml of cells.
11. Add 1/100 volume of EGTA, wait 5 minutes, and transform as in Steps 6 and 7.

Transformation and Storage of Competent Yeast Cells*

Materials

LAG Solution: 0.1 M lithium acetate, 15% glycerol in TE Buffer.

TE Buffer: 10 mM Tris-HCl pH 7.6, 0.1 mM EDTA.

LA Solution: 0.1 M lithium acetate in TE Buffer.

PEG Solution: 50% polyethylene glycol 4000.

YNBD Medium: 2 agar plates.

YEPD Medium: 60 ml.

Plasmid DNA: 0.1 to 1 μg of pEP100.

Note: The YNBD minimal medium should be supplemented with amino acids according to the genotype of the yeast auxotroph and the plasmid vectors. For this protocol, the shuttle plasmid vector pEP100 (Ap^r and Leu-2$^+$) and yeast strain 21D (mat α, leu2-3, leu2-112, his4, lys1) have been used. Plasmid DNA prepared using the CsCl, Triton cleared lysate and alkaline-SDS procedures can be used to transform yeast with this method. The frequency of transformation is strain-dependent and may vary as much as 10-fold from one strain to the next.

---

*This protocol is a modification of the one reported by Ito et al. (1983). J. Bacteriol. 153, 163. The yeast strain 21D was kindly provided by J. D. Cohen.

Protocol

1. Grow yeast to stationary phase of growth in 5 ml of YEPD medium. Use 0.05 ml of this culture to start a fresh 50 ml culture. Incubate this culture at 30°C with aeration, until an A_{600} of 1.5 to 2 is reached.

2. Harvest cells by centrifugation in a tabletop centrifuge and wash the cell pellet twice with 10 ml of TE Buffer.

3. Resuspend the cells in 5 ml of LA solution and continue the incubation (30°C with aeration) for 1 hour.

4. Harvest the competent cells by centrifugation and resuspend the cells in an equal volume of LAG solution. Dispense 0.3 ml aliquots of the cell suspension into sterile microfuge tubes and store at −70°C for up to 2 months.

5. To 0.3 ml of competent cells, add 10 µl of plasmid DNA (0.1 to 10 µg). Add 10 µl of TE Buffer to another tube of competent cells to serve as the plasmid-free control. Add 0.7 ml of PEG solution to each tube and invert three times. Incubate the transformation mixture at 30°C for 1 hour.

6. Following this incubation, heat the cells to 42°C for 5 minutes and then plate 0.1 ml of the mixture directly onto a YNBD plate. Plate 0.1 ml of the plasmid-free control mixture on the other YNBD plate.

7. Incubate the plates for 3 to 5 days at 30°C.

Assay for Chloramphenicol Acetyltransferase (CAT) Activity*

The chloramphenicol acetyltransferase (CAT) assay can be divided into three steps: preparation of the crude cell extract, quantitation of CAT activity for each sample, and determination of the protein concentration of the crude extract. Once those three steps are accomplished, the specific activity of CAT can be calculated.

A sonicator and scanning spectrophotometer equipped with a constant temperature water circulator for the sample chamber are the only special equipment required for the CAT assay.

Materials

Preparaton of Crude Cell Extract

    TDTT Buffer: 50 mM Tris-HCl, pH 7.8, 30 µM DL-dithiothreitol (Cleland's reagent).

    0.45 mm Glass Beads or Sonicator.

    For *E. coli*:

        LB medium supplemented with 0.2% glucose.

---

*Adapted from the CAT assay described by W. V. Shaw (1975). Meth. Enzymol. 43, 737.

APPENDIX 2

PROTOCOLS

LB agar plate containing 20 µg/ml ampicillin.

LB agar plates containing any antibiotics necessary to verify plasmid phenotype.

For yeast:

Competent yeast cells.

Appropriate selective plates for isolation of yeast transformants—these plates are determined by the phenotype of the plasmids to be examined.

Yeast minimal medium.

Protocol

Preparation of Crude Extracts from E. coli *for chloramphenicol Acetyltransferase Assays*

DAY 1

1. Make fresh streaks of all strains to be assayed on a selective antibiotic or minimal plate(s). Include a fresh streak of a strain containing pBR328 as a control for reagents and solutions used in the assay. Freshly streaked plates are important to ensure consistent results.

DAY 2

1. Inoculate 2 ml of LB medium supplemented with 0.2% glucose with a fresh colony and grow overnight at 37°C.

2. Inoculate 3 ml of LB supplemented with 0.2% glucose with 30–100 µl of stationary culture.

3. Incubate cultures 1 to 3 hours at 37°C.

4. When the A_{450} is 1.2, pellet the cells in a tabletop centrifuge, or Sorvall HS-4 rotor, 4.5 krpm, 8 minutes. (Low speed is used to prevent the culture tubes from breaking.)

5. Resuspend the cell pellets in 1 ml of TDTT Buffer, vortex vigorously, and transfer cell suspension to a microfuge tube. *Keep all samples on ice.*

6. Pellet the cells in a microfuge (spin 1 minute). Resuspend the samples in 300 µl TDTT Buffer.

7. Sonicate to lyse the cells. Use a microtip, and set the sonicator for the maximum level of the microtip. Sonicate the samples twice with a 15-second pulse or until the sample is clear. The samples should be kept on ice while they are sonicated. The cells may also be lysed by adding 0.45 mm glass beads to the meniscus, and vortexing the samples for 45 seconds. In this case, add 100λ additional TDTT Buffer.

8. Spin the samples in a microfuge for 15 minutes at 4°C.

9. Transfer the supernatants to clean microfuge tubes and keep on ice until assayed. If the CAT assay cannot be done on the same day, freeze the samples at −20°C until it is convenient to perform the assay. The samples will retain CAT activity for at least 1 month.

Preparation of Crude Extracts from Yeast for CAT Assays

APPENDIX 2
PROTOCOLS

DAY 3

1. Make fresh transformation into yeast with pEP100 and pEP101. These plasmids have substantial loss of CAT activity when stored on plates 2 to 3 weeks. Therefore for consistent results, fresh transformants are required. pEP100 and pEP101 are controls for your yeast crude extract preparation.

2. Make fresh transformations or streaks on selective plates for all yeast clones to be assayed.

DAY 4

1. Make a fresh streak of an *E. coli* strain containing pBR328 on an LB agar plate supplemented with 20 µg/ml ampicillin. This is the positive control of reagents and solutions used in the assay.

DAY 5

1. Treat pBR328 control according to protocol for preparation of *E. coli* crude extracts.

2. Inoculate 17 ml of yeast minimal media with 0.3 ml of stationary yeast culture and incubate overnight at 30°C.

   Note: The cells grow slowly. In the morning the A_{600} should be close to 1.5, i.e., log phase. If A_{600} is 1.5–2.0, go on to the next step. If the O.D. is too high, dilute the sample to an O.D. of 0.40 and grow for 3 to 5 hours. When measuring the O.D., dilute all cultures 1:1. This will take all samples below an O.D. of 1.0 and will increase the accuracy of the reading.

3. When an A_{600} of 1.5 is obtained, transfer the cells to 50 ml polypropylene centrifuge tubes and centrifuge the cells in the Sorvall HS-4 rotor 6.5 krpm for 7 to 10 minutes.

4. Add 1 ml of TDTT Buffer, resuspend the cell pellets, and transfer samples to microfuge tubes.

5. Centrifuge samples 1 minute in a microfuge. If the yeast samples are growing asynchronously, overlay the pellets with 300 µl of TDTT Buffer and store frozen until all of the samples have reached the correct O.D.

6. Resuspend the cell pellets in the 300 µl of TDTT Buffer.

7. Add acid washed glass beads to a level slightly below the meniscus and vortex at high speed 3X for 15 seconds. Add 100λ TDTT Buffer.

8. Spin samples 15 minutes in the cold, and transfer the supernatants to clean microfuge tubes. Keep samples on ice until assayed.

9. Follow standard CAT Assay procedure.

APPENDIX 2

PROTOCOLS

Materials

*Chloramphenicol Acetyltransferase Assay**

Crude Extracts.

5 mM Acetyl-CoA: 20 mg Acetyl-CoA (lithium salt) in 5 ml ddH$_2$O (make 0.2 ml aliquots in microfuge tubes and store frozen up to 1 year).

10 µM DTNB: 40 mg 5,5′-dithiobis-(2 nitrobenzoic acid) in 10 ml 1 M Tris-HCl pH 7.8 (*Make fresh each time*).

5 mM Cm: 32 mg chloramphenicol in 20 ml ddH$_2$O.

Protocol

1. Prepare reaction mixture for CAT assay as follows:

   8.8 ml ddH$_2$O
   1.0 ml 10 µM DTNB
   0.2 ml 5 mM Acetyl-CoA
   10.0 ml Total

   Keep the reaction mixture on ice. (10.0 ml of reaction mixture is sufficient for approximately 8 samples.)

2. Start constant temperature water circulator and warm sample chamber to 37°C.

3. For Beckman Model 24, set chart recorder span to 0.5 and rate at 1 inch/minute.

4. Add 0.588 ml reaction mixture to reference cuvette and sample cuvette. Let reaction mixtures warm up approximately 2 minutes. Adjust A$_{412}$ to 0.0.

5. Add 20 µl of crude extract to the sample cuvette, mix well, and let the sample equilibrate for 3 minutes.

6. Start the recorder to obtain background rate of acetylation. In approximately 3 minutes the slope becomes constant. When a constant slope is obtained, turn off the recorder drive.

7. Add 12 µl 5 mM Cm and mix well. Record the A$_{412}$ for 3 to 4 minutes. If the charge recorder needle moves quickly off the paper, CAT activity is high. A more accurate level of activity can be obtained by diluting the sample or changing the chart recorder span or rate.

8. The crude extracts can be stored frozen until protein concentration is determined.

9. Calculation of CAT activity
   a. Determine the slopes (ΔA_{412}/min.) before and after adding Cm.
   b. Subtract the background slope from the sample slope.

---

*Ibid.

c. Divide by the extinction coefficient 0.0136 to obtain nmoles/min.
d. Adjust for the dilution and volume of crude extract used in the assay to determine nmoles/min.ml. Example:

$$(\Delta OD)(\frac{1}{0.0136})(\frac{620\lambda}{20\lambda}) = \text{nmole/min.ml.}$$

Materials

Determination of Protein Concentrations and Calculation of Specific Activity

Bovine Serum Albumin.

Crude Extracts.

Biorad Protein Assay Dye Reagent Concentrate (Catalog #500-0006).

Protocol

1. Dissolve 5 mg BSA in 15 ml ddH$_2$O (solution A).
2. Read A$_{280}$ of BSA solution A against a ddH$_2$O blank. Adjust protein concentration to A$_{280}$ = 0.165 (250 µg/ml BSA).
3. Dilute 1 ml of BSA solution A with 9 ml ddH$_2$O. (Final concentration: 25 µg BSA/ml) and label bottle as solution B.
4. Using solution B make the following dilutions and dye reagent additions:

Sol. B	ddH$_2$O	dye reagent	µg protein
0.8 ml	0.0 ml	0.2 ml	20.0
0.7	0.1	0.2	17.5
0.6	0.2	0.2	15.0
0.5	0.3	0.2	12.5
0.4	0.4	0.2	10.0
0.3	0.5	0.2	7.5
0.2	0.6	0.2	5.0
0.1	0.7	0.2	0.0
*0.0	0.8	0.2	0.0
*0.0	0.8	0.2	0.0

*Samples for reference and blank.

5. Mix the solutions gently and let stand 5 minutes to 1 hour.
6. Read A$_{595}$. Plot a graph of µg of protein vs. A$_{595}$. Calculate µg protein/A$_{595}$.
7. For each sample that will be assayed, add 0.8 ml ddH$_2$O to a small culture tube or microfuge tube. Also add 0.8 ml ddH$_2$O to two tubes for reference and blank.

APPENDIX 2
PROTOCOLS

8. Add 0.2 ml dye reagent to each of the above samples.
9. Add 5–10 μl of crude extract.
10. Gently mix the samples and let them stand 5 minutes to 1 hour.
11. Read A_{595} against the blank.
12. Calculate the protein concentration of crude extract:

$$(A_{595})(\mu g\ BSA/A_{595})\left(\frac{1}{\text{volume assayed (5–10 }\mu\text{l)}}\right) = \mu g/\mu l \text{ or } mg/ml$$

13. Calculate the specific activity of CAT in samples:

$$(n\ mole/min.ml)\left(\frac{1}{mg/ml}\right) = n\ mole/min.mg.$$

Maxicell Purification and Labeling of Proteins

Materials

E. coli CSR603 F⁻ *thr*-1 *leu*B6 *pro*A2 *phr*-1 *rec*A1 *arg*E3 *thi*-1 *uvr*A6 *ara*-14 *lac*Y1 *gal*K2 *xyl*-5 *mtl*-1 *rps*L31 *tsx*-33 λ⁻ *sup*E44 (mucoid strain).

LB medium.

K Medium: M-9 glucose medium containing 1% casamino acids + 0.1 μg/ml thiamine hydrochloride.*

Hershey Salts*:

NaCl	5.4 gm
KCl	3.0 gm
NH₄Cl	1.1 gm
CaCl₂·2H₂O	15 mg
MgCl₂·6H₂O	200 mg
FeCl₃·6H₂O	0.2 mg
KH₂PO₄	87 mg
Tris base	12.1 gm
ddH₂O	800 ml

Adjust pH to 7.4 by the addition of concentrated HCl.

ddH₂O to	1 liter volume

Hershey Medium for CSR603:†

Hershey salts	100 ml
20% glucose	2 ml
2% threonine	0.5 ml
1% leucine	1 ml
2% proline	1 ml
2% arginine	1 ml
0.1% thiamine hydrochloride	0.1 ml

*Sancar et al. (1979). J. Bacteriol. 137, 692; Tait et. al. (1982) Gene 20, 39.
†Ibid.

APPENDIX 2
PROTOCOLS

15 mg/ml D-cycloserine.

^{35}S-methionine (~ 1000 Ci/mMole).

100 mM NaCl.

Sample Buffer (1X):

SDS (sodium dodecyl sulfate, British Drug House, Ltd.)	2.3 gm
2-mercaptoethanol	5 ml
glycerol	10 ml
Bromophenol blue	0.05 gm
1 M Tris-HCl pH 6.8	6.25 ml
ddH$_2$O to	100 ml volume

Protocol

DAY 1

1. Prepare competent CaCl$_2$-treated CSR603 and transform with the plasmids to be analyzed. Plate on appropriate selective media and incubate at 37°C.

DAYS 2 AND 3

1. Grow 2 ml LB cultures of transformants obtained from Day 1. Use a miniscreen procedure to isolate plasmid DNA and verify the presence of the plasmids of interest.

DAY 4

1. After the presence of the plasmids in CSR603 has been verified, proceed with the preparation and labeling of maxicells.

2. Inoculate 1 ml of K medium with CSR603 containing no plasmid and with CSR603 containing each of the plasmids of interest. Incubate cultures at 37°C with shaking overnight.

DAY 5

1. Dilute 0.1 ml of each stationary culture into 2.9 ml K medium. Incubate at 37°C with shaking to 2×10^8 cells/ml ($A_{600} = 0.5$).

2. Transfer 2.5 ml of each culture to a petri dish. Remove covers and irradiate cells for 15 seconds with a UV dose of ~ 50 J/m^2.

3. Transfer 2 ml of the irradiated cells to a test tube containing 15 μl of 15 mg/ml cycloserine. Incubate at 37°C with shaking for 16 to 20 hours.

DAY 6

1. Place 1.5 ml of the incubated, irradiated cells in a microfuge tube. Centrifuge 2 minutes at room temperature. Pour off the supernatants.

2. Wash cell pellets twice with 1 ml of Hershey salts and resuspend in 0.8 ml of Hershey medium.

3. Incubate 1 hour at 37°C, then add 0.2 ml of Hershey medium containing 5 μCi ^{35}S-methionine. Incubate 2 hours at 37°C.

4. Pellet the labeled cells and wash twice with 1 ml of 100 mM NaCl. The supernatants from each centrifugation should be treated as radioactive waste.

5. Resuspend each cell pellet in 50 μl of sample buffer. Vortex samples, then heat at 100°C for 3 to 5 minutes to lyse cells and denature proteins.

6. Subject 15 μl samples to electrophoresis on 12.5% SDS-polyacrylamide protein gels.

7. Detect the radioactive plasmid-encoded polypeptides by autoradiography or fluorography.

Minicell Purification and Labeling of Proteins

Materials

Minicell-Producing Strains of *E. coli*:

P678-54 (F$^-$ *thr$^-$ leu$^-$ thi$^-$ supE lacY tonA gal$^-$ mal$^-$ xyl$^-$ ara$^-$ mtl$^-$ min$^-$*)

X1488 (F$^-$ *str^r hst$^-$ hsm$^+$ minA$^-$ minB$^-$ purE$^-$ pdxC his$^-$ ile$^-$ met$^-$ ade$^-$ ura$^-$* r$_k^-$m$_k^+$)

Plasmid DNA for Transformations.

Reagents for Production of Competent Cells: Exercise 9.

Reagents for Miniscreen DNA Isolation: Exercise 5.

M9 Growth Medium: M9 medium + 0.4% glucose + 0.4% casamino acids + any nutritional supplements required by genotype of minicell producing strain (i.e., supplemented with *thr, leu,* and B$_1$ for P678-54).

M9 Labeling Medium: M9 medium + 0.4% glucose + any nutritional supplements required by the strain. *Omit* casamino acids and methionine.

BSG Buffer:

NaCl	4.25 gm
KH$_2$PO$_4$	0.15 gm
Na$_2$HPO$_4$	0.30 gm
gelatin	50 mg
ddH$_2$O to	500 ml volume

Autoclave to dissolve gelatin.

Sucrose Solutions: 50 mM Tris-HCl pH 7.5, 100 mM NaCl, 0.1 mM EDTA, 5%, 12.5%, and 20% sucrose.

TE Buffer: 10 mM Tris-HCl pH 7.6, 1 mM Na$_2$EDTA.

Sample Buffer (1X): 10% glycerol, 5% 2-mercaptoethanol, 2.3% sodium dodecyl sulfate (British Drug House, Ltd.), 0.625 M Tris-HCl pH 6.8, .01% Bromophenol blue (see protocol for denaturing protein gels).

1 mg/ml Cycloserine.

Protocol

DAY 1

1. Prepare competent cells of the minicell-producing strain. Transform with the plasmid DNAs to be examined. Plate on appropriate selective agar plates and incubate at 37°C.

DAYS 2 AND 3

1. Prepare 2 ml LB cultures of the transformants. Isolate DNA using a miniscreen procedure. Verify the presence of the desired plasmids.

2. At approximately 4:00 to 5:00 in the afternoon, use a fresh bacterial colony to inoculate a 250 ml of M9 growth medium in a 500 ml culture flask with the minicell strain (as a control) and with each of the plasmid-containing transformants. Six different cultures can be conveniently processed simultaneously. Incubate 16 hours at 37°C with shaking.

DAY 4

1. Build 2 sucrose gradients for *each* minicell culture to be processed. These will be used in the Beckman SW27 rotor and should be built in clear 1" diameter by 3½" tubes. Gradients may be step gradients consisting of 12 ml each of 20%, 12.5%, and 5% sucrose, or linear 20-5% sucrose gradients. Linear gradients can be established by layering 18 ml of 5% sucrose on 18 ml of 20% sucrose, then subjecting to three cycles of freezing and thawing.

2. While preparing the sucrose gradients, clear the minicell cultures of viable cells by low-speed centrifugation (2.5 krpm, 4°C, 10 minutes, Sorvall GSA rotor). Save *supernatants* and discard pellets.

3. Harvest minicells by high-speed centrifugation (8.5 krpm, 4°C, 15 minutes, Sorvall GSA rotor). Discard supernatants and save *minicell pellets*.

4. Resuspend minicell pellets in 2 ml cold BSG. Layer each sample onto a sucrose gradient and centrifuge (5 krpm, 4°C, 15 minutes, Beckman SW27 rotor).

5. Gradients will contain a viable cell pellet and a broad minicell band. The location of this band may vary with different minicell-producing strains, and some strains will produce additional bands of small viable cells. Particular care should be taken when withdrawing the minicell band, usually located at the interface of the 5% and 12.5% sucrose layers. Use a pasteur pipet to transfer this band to a 50 ml polypropylene tube.

APPENDIX 2
PROTOCOLS

6. Add an equal volume of cold BSG. Invert tubes to mix. Pellet minicells by centrifugation (12 krpm, 4°C, 15 minutes, Sorval SS34 rotor).

7. Resuspend each minicell pellet in 10 ml M9 growth medium containing 1 µg/ml cycloserine. Incubate at 37°C with shaking for 60 minutes. The cycloserine will cause the lysis of any actively growing viable cells in the preparation of minicells.

8. Concentrate minicells by centrifugation (12 krpm, 4°C, 15 minutes, Sorvall SS34 rotor). Resuspend in 2 ml cold BSG.

9. Layer each minicell sample onto a second sucrose gradient and repeat Steps 4, 5, and 6.

10. Resuspend the resulting (somewhat fragile) minicell pellet in 1 ml of M9 Labeling Medium and adjust A_{620} to $0.2 \pm .02$.

11. Plate 0.1 ml of a 10^{-1} dilution of each minicell preparation on an LB agar plate. Incubate at 37°C. This plating will indicate the level of viable cell contamination. A contamination level of greater than 10^5 viable cells/ml will interfere with detection of minicell-directed polypeptide synthesis.

12. Place 1.0 ml of each minicell sample in a sterile 15 ml Corex tube. Preincubate 15 minutes at 37°C with shaking.

13. To each sample add 5–50 µCi ^{35}S-methionine. Continue 37°C incubation for 30 minutes.

14. Harvest labeled minicells by centrifugation (12 krpm, 4°C, 15 minutes, Sorvall SS34 rotor). Remove the supernatants with a pasteur pipet as soon as the rotor has come to a stop. If allowed to stand too long, the minicell pellets will rapidly resuspend. Supernatants and pipets should be considered radioactive waste.

15. Resuspend minicell pellets in 1 ml cold TE Buffer. Concentrate by centrifugation and remove the (radioactive) wash supernatants.

16. Add 40 µl Sample Buffer. Vortex 60 seconds to resuspend minicells. Centrifuge in a tabletop centrifuge 2 minutes, then withdraw the samples with a manual pipetting device and place in labeled microfuge tubes.

17. Tightly cap tubes and place at 100°C for 3 minutes. Minicells will lyse and the samples will change from cloudy to clear blue.

18. Load 5–15 µl samples on 12.5% SDS-polyacrylamide denaturing protein gels and subject to electrophoresis.

19. Following electrophoresis, radioactive polypeptides can be detected by autoradiography or fluorography.

References

Adler et al. 1967. Proc. Natl. Acad. Sci. USA 57:321.
Roozen et al. 1971. J. Bacteriol. 107:21.
Meagher et al. 1977. Cell 10:521.

SDS-polyacrylamide Denaturing Protein Gels

APPENDIX 2
PROTOCOLS

Materials

Acrylamide.

N,N'-Methylene-bis-Acrylamide.

TEMED (N,N,N',N'-tetramethylethylenediamine).

Trizma Base.

HCl: 1N.

SDS (sodium dodecyl sulfate, BDH Chemicals Ltd.).

Ammonium Persulfate.

Glycine.

Glycerol.

2-mercaptoethanol (β-mercaptoethanol).

Bromophenol Blue.

Acetic Acid: Glacial.

Methanol.

Coomassie Brilliant Blue R.

Note: These gels consist of a lower resolving gel and an upper stacking gel that concentrates the sample before its entry into the resolving gel. The buffer system is the discontinuous system of Laemmli as described by O'Farrell.

Lower Gel Buffer:*

    1.5 M TRIS-HCl pH 8.8.

    0.4% SDS.

Upper Gel Buffer:†

    0.5 M TRIS-HCl pH 6.8.

    0.4% SDS.

Acrylamide Stock:‡

    29.2% acrylamide.

    0.8% N,N'-methylene-bis-acrylamide.

    ddH_2O to 100 ml volume.

    Dissolve, filter to remove particulate matter, and store at 4°C.

10% Ammonium persulfate (make fresh every time):

    0.1 gm. ammonium persulfate

    0.95 ml ddH_2O

---

*O'Farrell (1975). J. Biol. Chem. 250:4007.
†Ibid.
‡Ibid.

APPENDIX 2

PROTOCOLS

The volumes below are for slab gels (152 × 152 × 0.8 mm) of 15 ml volume with a 6 ml stacking gel. Gel solutions should be mixed and degassed under vacuum prior to the initiation of polymerization by the addition of TEMED. Volumes are given in ml.

Reagent	Resolving Gel						4% Stacking Gel
	5%	7.5%	10%	12.5%	15%	17.5%	
1. Lower Gel Buffer	3.75	3.75	3.75	3.75	3.75	3.75	---
2. Upper Gel Buffer	---	---	---	---	---	---	1.25
3. Acrylamide Stock	2.50	3.75	5.00	6.25	7.50	8.75	.80
4. ddH$_2$O	8.7	7.45	6.20	4.95	3.70	2.45	3.93
5. 10% ammonium persulfate	.05	.05	.05	.05	.05	.05	.02
	De-gas						De-gas
6. TEMED	.007	.007	.007	.007	.007	.007	.004

Protocol

1. The gel plates should be assembled and sealed against leakage (see fig. 5.4a–c). The resolving gel can be poured and covered with a 2–3 mm layer of butanol to exclude air and form an even polymerization front.

2. After polymerization, the butanol can be removed by flushing briefly with water, then the stacking gel poured and a comb inserted to form sample wells.

3. Following polymerization of the stacking gel, the comb is removed and the wells rinsed to remove unpolymerized acrylamide.

4. The gel is then assembled in the running apparatus and running buffer added.

   SDS Running Buffer:*

   0.025 M Trizma base.

   0.192 M glycine.

   0.1% SDS.

5. Protein samples are prepared by adding an equal volume of 2 × SDS Sample Buffer and heating at 98–100°C for 3 to 5 minutes to ensure denaturation of the samples.

   2 × SDS Sample Buffer:

   20% (w/v) glycerol.

   10% (w/v) 2-mercaptoethanol.

   4.6% (w/v) SDS.

   0.125 M Tris-HCl pH 6.8.

   0.1% Bromophenol blue.

*Ibid.

6. Gels are typically run at 5–40 milliamps (mA) for 2 to 8 hours or until the marker dye reaches the end of the gel.

7. The gel apparatus is dissembled and the gel removed. The gel can be immediately dried on Whatman 3MM filter paper for autoradiography, or stained for direct visualization of polypeptide bands.

Note: A variety of staining solutions and protocols are available, many using the stain Coomassie Brilliant Blue. The procedure below gives rapid, reproducible visualization of bands in 1 to 2 hours.

Staining Solution:

5% (v/v) acetic acid.

10% (v/v) methanol.

Stain:

1% Coomassie Brilliant Blue in ddH_2O.

Destaining Solution:

7% (v/v) acetic acid.

Protocol

1. Place gel in a staining container with lid. Add 50 ml staining solution. Cover and place at 50–55°C for 15 minutes. This wash helps fix the polypeptide bands and removes SDS from the gel to allow more efficient staining.

2. Pour off solution, replace with 50 ml Staining Solution + 5 ml Stain. Cover and place at 50–55°C for 15 to 20 minutes. Gel will turn blue at this time.

3. Pour off the staining solution. Add 100 ml Destaining Solution, cover and return to 50–55°C for 15 minutes. Gentle mixing will improve the destaining process.

4. Pour off the destaining solution and replace.

5. Continue to repeat Steps 3 and 4. Polypeptide bands will generally become visible after 5 to 15 minutes of destaining. Complete destaining to produce a clear gel with blue polypeptide bands depends on gel thickness and may require 2 to 24 hours. Remember that repeated small volume washes will destain more effectively than extended large-volume washes. Wash periods can be extended for convenience, and temperatures from 20°–65°C are suitable for the procedure.

References

Laemmli. 1970. Nature 227:680.
O'Farrell. 1975. J. Biol Chem 250:4007.

APPENDIX 2
PROTOCOLS

Preparation of Dialysis Tubing and Minidialysis Chamber

Material

Dialysis Tubing: 1-inch width.

Disposable Rubber Gloves.

Sodium Bicarbonate ($NaHCO_3$).

TE Buffer: 10 mM Tris-HCl pH 7.6, 1 mM EDTA.

Ethanol: 95%.

Pipetman tip: 1ml.

Amber Rubber Tubing: ¼ inch (internal diameter).

Protocol (Dialysis tubing)

1. Rinse 20 feet of dialysis tubing in 2 liters of ddH_2O in a 4-liter Erlenmeyer flask.

2. Add 50 gm of $NaHCO_3$, mix, and boil or autoclave for 15 minutes.

3. After the solution has cooled, pour off the $NaHCO_3$. Wearing rubber gloves, strip the dialysis tubing (between the thumb and forefinger) into 3 liters of ddH_2O.

4. Allow the tubing to soak for 5 minutes and pour off the water. Repeat Steps 3 and 4.

5. Now strip the tubing into 2 liters of TE Buffer and boil or autoclave for 10 minutes.

6. Let the solution cool and add an equal volume of 95% EtOH. Store at 4°C.

Protocol (Minidialysis chamber)

1. As shown in Fig. A2.1a, cut a 1 ml Pipetman tip 28 mm from the small end. Fire polish the cut end by quickly passing it through the flame of a Bunsen burner.

2. Cut a ½-inch piece of amber rubber tubing. Wearing rubber gloves, use stainless steel scissors (rinsed in ethanol) to cut a 1-inch piece of dialysis tubing. Next, cut the piece of dialysis tubing along one edge.

Figure A2.1

3. Unfold the dialysis tubing and place it over the cut end of the Pipetman tip. Secure the dialysis tubing by sliding the piece of amber tubing over the end (fig. A2.1b).

4. Tape the dialysis chamber to the inside edge of a 1-liter beaker containing Dialysis Buffer. Place the chamber about 2 mm below the surface of the buffer (fig. A2.1c).

5. Add the sample to be dialyzed into the chamber and dialysis at 4°C (with stirring) for at least 4 hours. A commonly used Dialysis Buffer is 10 mM Tris-HCl pH 7.6, 10 mM NaCl, 1 mM EDTA (TEN Buffer).

APPENDIX 3

Restriction Maps and Nucleotide Sequences

APPENDIX 3

RESTRICTION MAPS AND NUCLEOTIDE SEQUENCES

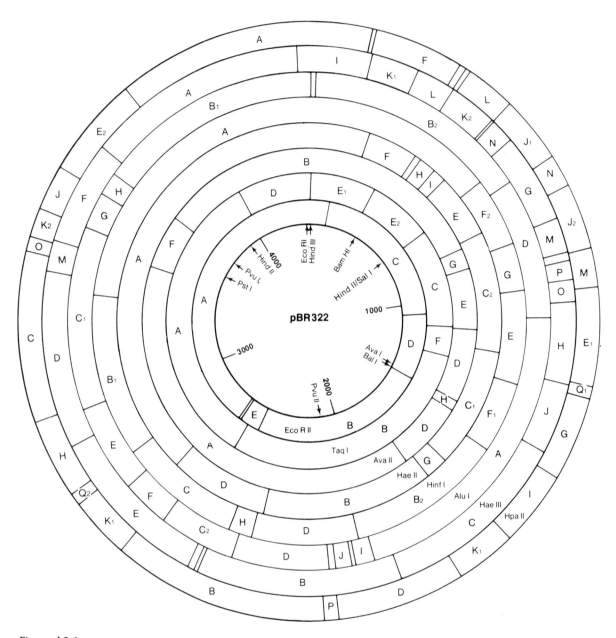

Figure A3.1
Restriction endonuclease cleavage map of pBR322. The data presented in this map were prepared from primary sequence data (J. G. Sutcliffe (1978) *Nucleic Acids Research* 5, 2721) by R. Blakesley, Bethesda Research Laboratories (Copyright, Bethesda Research Laboratories, Inc., 1981). Restriction enzymes that do not cleave pBR322 are: *Ava*X, *Bcl*I, *Bgl*II, *Eca*I, *Kpn*I, *Sma*I, *Sst*I, *Xba*I, and *Xho*I.

                    Eco RI   HindIII
                                   Tc^r promoter
    (GAA)TTCTCATGTTTGACAGCTTATCATCGATAAGCTTTAATGCGGTAGTTTATCACAGTTAAATTGCTAACGCAGTCAGGCACCGTGTATGAAATCTAACAAT
                                                                                                    100.
    AAGAGTACAAACTGTCGAATAGTAGCTATTCGAAATTACGCCATCAAATAGTGTCAATTTAACGATTGCGTCAGTCCGTGGCACATACTTTAGATTGTTA
    ←∼∼∼∼∼∼∼∼∼∼ α-Tc^r promoter
    GCGCTCATCGTCATCCTCGGCACCGTCACCCTGGATGCTGTAGGCATAGGCTTGGTTATGCCGGTACTGCCGGGCCTCTTGCGGGATATCGTCCATTCCG
                                                                                                    200.
    CGCGAGTAGCAGTAGGAGCCGTGGCAGTGGGACCTACGACATCCGTATCCGAACCAATACGGCCATGACGGCCCGGAGAACGCCCTATAGCAGGTAAGGC

    ACAGCATCGCCAGTCACTATGGCGTGCTGCTAGCGCTATATGCGTTGATGCAATTTCTATGCGCACCCGTTCTCGGAGCACTGTCCGACCGCTTTGGCCG
                                                                                                    300.
    TGTCGTAGCGGTCAGTGATACCGCACGACGATCGCGATATACGCAACTACGTTAAAGATACGCGTGGGCAAGAGCCTCGTGACAGGCTGGCGAAACCGGC

                                                                                        BamHI
    CCGCCCAGTCCTGCTCGCTTCGCTACTTGGAGCCACTATCGACTACGCGATCATGGCGACCACACCCGTCCTGTGGATCCTCTACGCCGGACGCATCGTG
                                                                                                    400.
    GGCGGGTCAGGACGAGCGAAGCGATGAACCTCGGTGATAGCTGATGCGCTAGTACCGCTGGTGTGGGCAGGACACCTAGGAGATGCGGCCTGCGTAGCAC

    GCCGGCATCACCGGCGCCACAGGTGCGGTTGCTGGCGCCTATATCGCCGACATCACCGATGGGGAAGATCGGGCTCGCCACTTCGGGCTCATGAGCGCTT
                                                                                                    500.
    CGGCCGTAGTGGCCGCGGTGTCCACGCCAACGACCGCGGATATAGCGGCTGTAGTGGCTACCCCTTCTAGCCCGAGCGGTGAAGCCCGAGTACTCGCGAA

                                                 v   SphI
    GTTTCGGCGTGGGTATGGTGGCAGGCCCCGTGGCCGGGGGACTGTTGGGCGCCATCTCCTTGCATGCACCATTCCTTGCGGCGGCGGTGCTCAACGGCCTC
                                                                                                    600.
    CAAAGCCGCACCCATACCACCGTCCGGGCACCGGCCCCCTGACAACCCGCGGTAGAGGAACGTACGTGGTAAGGAACGCCGCCGCCACGAGTTGCCGGAG
                                                 ^   SalI
    AACCTACTACTGGGCTGCTTCCTAATGCAGGAGTCGCATAAGGGAGAGCTCGACCGATGCCCTTGAGAGCCTTCAACCCAGTCAGCTCCTTCCGGTGGG
                                                                                                    700.
    TTGGATGATGACCCGACGAAGGATTACGTCCTCAGCGTATTCCCTCTCGCAGCTGGCTACGGGAACTCTCGGAAGTTGGGTCAGTCGAGGAAGGCCACCC

    CGCGGGGCATGACTATCGTCGCCGCACTTATGACTGTCTTCTTTATCATGCAACTCGTAGGACAGGTGCCGGCAGCGCTCTGGGTCATTTTCGGCGAGGA
                                                                                                    800.
    GCGCCCCGTACTGATAGCAGCGGCGTGAATACTGACAGAAGAAATAGTACGTTGAGCATCCTGTCCACGGCCGTCGCGAGACCCAGTAAAAGCCGCTCCT

    CCGCTTTCGCTGGAGCGCGACGATGATCGGCCTGTCGCTTGCGGTATTCGGAATCTTGCACGCCCTCGCTCAAGCCTTCGTCACTGGTCCCGCCACCAAA
                                                                                                    900.
    GGCGAAAGCGACCTCGCGCTGCTACTAGCCGGACAGCGAACGCCATAAGCCTTAGAACGTGCGGGAGCGAGTTCGGAAGCAGTGACCAGGGCGGTGGTTT

    CGTTTCGGCGAGAAGCAGGCCATTATCGCCGGCATGGCGGCCGACGCGCTGGGCTACGTCTTGCTGGCGTTCGCGACGCGAGGCTGGATGGCCTTCCCCA
                                                                                                    1000.
    GCAAAGCCGCTCTTCGTCCGGTAATAGCGGCCGTACCGCCGGCTGCGCGACCCGATGCAGAACGACCGCAAGCGCTGCGCTCCGACCTACCGGAAGGGGT

    TTATGATTCTTCTCGCTTCCGGCGGCATCGGGATGCCCGCGTTGCAGGCCATGCTGTCCAGGCAGGTAGATGACGACCATCAGGGACAGCTTCAAGGATC
                                                                                                    1100.
    AATACTAAGAAGAGCGAAGGCCGCCGTAGCCCTACGGGCGCAACGTCCGGTACGACAGGTCCGTCCATCTACTGCTGGTAGTCCCTGTCGAAGTTCCTAG

    GCTCGCGGCTCTTACCAGCCTAACTTCGATCACTGGACCGCTGATCGTCACGGCGATTTATGCCGCCTCGGCGAGCACATGGAACGGGTTGGCATGGATT
                                                                                                    1200.
    CGAGCGCCGAGAATGGTCGGATTGAAGCTAGTGACCTGGCGACTAGCAGTGCCGCTAAATACGGCGGAGCCGCTCGTGTACCTTGCCCAACCGTACCTAA

    GTAGGCGCCGCCCTATACCTTGTCTGCCTCCCCGCGTTGCGTCGCGGTGCATGGAGCCGGGCCACCTCGACCTGAATGGAAGCCGGCGGCACCTCGCTAA
                                                                                                    1300.
    CATCCGCGGCGGGATATGGAACAGACGGAGGGGCGCAACGCAGCGCCACGTACCTCGGCCCGGTGGAGCTGGACTTACCTTCGGCCGCCGTGGAGCGATT

    CGGATTCACCACTCCAAGAATTGGAGCCAATCAATTCTTGCGGAGAACTGTGAATGCGCAAACCAACCCTTGGCAGAACATATCCATCGCGTCCGCCATC
                                                                                                    1400.
    GCCTAAGTGGTGAGGTTCTTAACCTCGGTTAGTTAAGAACGCCTCTTGACACTTACGCGTTTGGTTGGGAACCGTCTTGTATAGGTAGCGCAGGCGGTAG

               AvaI   BalI
    TCCAGCAGCCGCACGCGGCGCATCTCGGGCAGCGTTGGGTCCTGGCCACGGGTGCGCATGATCGTGCTCCTGTCGTTGAGGACCCGGCTAGGCTGGCGGG
                                                                                                    1500.
    AGGTCGTCGGCGTGCGCCGCGTAGAGCCCGTCGCAACCCAGGACCGGTGCCCACGCGTACTAGCACGAGGACAGCAACTCCTGGGCCGATCGACCGCCC

    GTTGCCTTACTGGTTAGCAGAATGAATCACCGATACGCGAGCGAACGTGAAGCGACTGCTGCTGCAAAACGTCTGCGACCTGAGCAACAACATGAATGGT
                                                                                                    1600.
    CAACGGAATGACCAATCGTCTTACTTAGTGGCTATGCGCTCGCTTGCACTTCGCTGACGACGACGTTTTGCAGACGCTGGACTCGTTGTTGTACTTACCA

    CTTCGGTTTCCGTGTTTCGTAAAGTCTGGAAACGCGGAAGTCAGCGCCCTGCACCATTATGTTCCGGATCTGCATCGCAGGATGCTGCTGGCTACCCTGT
                                                                                                    1700.
    GAAGCCAAAGGCACAAAGCATTTCAGACCTTTGCGCCTTCAGTCGCGGGACGTGGTAATACAAGGCCTAGACGTAGCGTCCTACGACGACCGATGGGACA

    GGAACACCTACATCTGTATTAACGAAGCGCTGGCATTGACCCTGAGTGATTTTTCTCTGGTCCCGCCGCATCCATACCGCCAGTTGTTTACCCTCACAAC
                                                                                                    1800.
    CCTTGTGGATGTAGACATAATTGCTTCGCGACCGTAACTGGGACTCACTAAAAAGAGACCAGGGCGGCGTAGGTATGGCGGTCAACAAATGGGAGTGTTG

    GTTCCAGTAACCGGGCATGTTCATCATCAGTAACCCGTATCGTGAGCATCCTCTCTCGTTTCATCGGTATCATTACCCCCATGAACAGAAATTCCCCCTT
                                                                                                    1900.
    CAAGGTCATTGGCCCGTACAAGTAGTAGTCATTGGGCATAGCACTCGTAGGAGAGAGCAAAGTAGCCATAGTAATGGGGGTACTTGTCTTTAAGGGGGAA

    ACACGGAGGCATCAAGTGACCAAACAGGAAAAAACCGCCCTTAACATGGCCCGCTTTATCAGAAGCCAGACATTAACGCTTCTGGAGAAACTCAACGAGC
                                                                                                    2000.
    TGTGCCTCCGTAGTTCACTGGTTTGTCCTTTTTTGGCGGGAATTGTACCGGGCGAAATAGTCTTCGGTCTGTAATTGCGAAGACCTCTTTGAGTTGCTCG
                                                                                              (continued)

Figure A3.2

Annotated nucleotide sequence of pBR322. The nucleotide sequence of pBR322 (4363 bp) is shown with most unique restriction sites indicated above the 5' strand. Also shown are the transcribed and translated regions of the plasmid. The origin and direction of DNA replication is indicated by "ori" and the C/G base pair omitted from the Sutcliffe sequence is indicated by "v" on both the 5' and 3' strands (34 bp to the left of the *Sph*I site). Annotation of the pBR322 sequence, as well as those shown for pBR327 and pBR329, was based on information obtained from the following references: J. G. Sutcliffe, Cold Spring Harb. Symp. Quant. Biol. 43:77, 1979 (nucleotide sequence of pBR322); K. W. C. Peden, Gene 22:275, 1983 (Tc^r gene and C/G omission); Chan et al., Nucl. Acid Res. 7:1247, 1979 (104 base RNA); Bolivar et al., Proc. Natl. Acad. Sci. USA 74:5265, 1977 (origin and direction of DNA replication); J. G. Sutcliffe, Proc. Natl. Acad. Sci. USA 75:3737, 1978 (Ap^r gene); Russel and Bennett, Nucl. Acid Res. 9:2517, 1981 and Brosius et al., J. Biol. Chem. 257:9205, 1982 (Ap^r promoter); Brosius et al., J. Biol. Chem. 257:9205 and West and Rodriguez, Gene 20:291, 1982 (Tc^r and α-Tc^r promoter); Alton and Vapnek, Nature 282:864, 1979 (nucleotide sequence of the Cm^r gene). (Nucleotide sequence courtesy of J. G. Sutcliffe.)

```
                                                                    Pvu II
TGGACGCGGATGAACAGGCAGACATCTGTGAATCGCTTCACGACCACGCTGATGAGCTTTACCGCAGCTGCCTCGCGCGTTTCGGTGATGACGGTGAAAA
                                                                                                  2100.
ACCTGCGCCTACTTGTCCGTCTGTAGACACTTAGCGAAGTGCTGGTGCGACTACTCGAAATGGCGTCGACGGAGCGCGCAAAGCCACTACTGCCACTTTT

CCTCTGACACATGCAGCTCCCGGAGACGGTCACAGCTTGTCTGTAAGCGGATGCCGGGAGCAGACAAGCCCGTCAGGGCGCGTCAGCGGGTGTTGGCGGG
                                                                                                  2200.
GGAGACTGTGTACGTCGAGGGCCTCTGCCAGTGTCGAACAGACATTCGCCTACGGCCCTCGTCTGTTCGGGCAGTCCCGCGCAGTCGCCCACAACCGCCC

TGTCGGGGCGCAGCCATGACCCAGTCACGTAGCGATAGCGGAGTGTATACTGGCTTAACTATGCGGCATCAGAGCAGATTGTACTGAGAGTGCACCATAT
                                                                                                  2300.
ACAGCCCCGCGTCGGTACTGGGTCAGTGCATCGCTATCGCCTCACATATGACCGAATTGATACGCCGTAGTCTCGTCTAACATGACTCTCACGTGGTATA

GCGGTGTGAAATACCGCACAGATGCGTAAGGAGAAAATACCGCATCAGGCGCTCTTCCGCTTCCTCGCTCACTGACTCGCTGCGCTCGGTCGTTCGGCTG
                                                                                                  2400.
CGCCACACTTTATGGCGTGTCTACGCATTCCTCTTTTATGGCGTAGTCCGCGAGAAGGCGAAGGAGCGAGTGACTGAGCGACGCGAGCCAGCAAGCCGAC

CGGCGAGCGGTATCAGCTCACTCAAAGGCGGTAATACGGTTATCCACAGAATCAGGGGATAACGCAGGAAAGAACATGTGAGCAAAAGGCCAGCAAAAGG
                                                                                                  2500.
GCCGCTCGCCATAGTCGAGTGAGTTCCGCCATTATGCCAATAGGTGTCTTAGTCCCCTATTGCGTCCTTTCTTGTACACTCGTTTTCCGGTCGTTTTCC
                                       ◀Ori
CCAGGAACCGTAAAAAGGCCGCGTTGCTGGCGTTTTTCCATAGGCTCCGCCCCCCTGACGAGCATCACAAAAATCGACGCTCAAGTCAGAGGTGGCGAAA
                                                                                                  2600.
GGTCCTTGGCATTTTTCCGGCGCAACGACCGCAAAAAGGTATCCGAGGCGGGGGGACTGCTCGTAGTGTTTTTAGCTGCGAGTTCAGTCTCCACCGCTTT

CCCGACAGGACTATAAAGATACCAGGCGTTTCCCCCTGGAAGCTCCCTCGTGCGCTCTCCTGTTCCGACCCTGCCGCTTACCGGATACCTGTCCGCCTTT
                                                                                                  2700.
GGGCTGTCCTGATATTTCTATGGTCCGCAAAGGGGGACCTTCGAGGGAGCACGCGAGAGGACAAGGCTGGGACGGCGAATGGCCTATGGACAGGCGGAAA

CTCCCTTCGGGAAGCGTGGCGCTTTCTCAATGCTCACGCTGTAGGTATCTCAGTTCGGTGTAGGTCGTTCGCTCCAAGCTGGGCTGTGTGCACGAACCCC
                                                                                                  2800.
GAGGGAAGCCCTTCGCACCGCGAAAGAGTTACGAGTGCGACATCCATAGAGTCAAGCCACATCCAGCAAGCGAGGTTCGACCCGACACACGTGCTTGGGG

CCGTTCAGCCCGACCGCTGCGCCTTATCCGGTAACTATCGTCTTGAGTCCAACCCGGTAAGACACGACTTATCGCCACTGGCAGCAGCCACTGGTAACAG
                                                                                                  2900.
GGCAAGTCGGGCTGGCGACGCGGAATAGGCCATTGATAGCAGAACTCAGGTTGGGCCATTCTGTGCTGAATAGCGGTGACCGTCGTCGGTGACCATTGTC
                                                            104 base RNA ~~~~~~~~~
GATTAGCAGAGCGAGGTATGTAGGCGGTGCTACAGAGTTCTTGAAGTGGTGGCCTAACTACGGCTACACTAGAAGGACAGTATTTGGTATCTGCGCTCTG
                                                                                                  3000.
CTAATCGTCTCGCTCCATACATCCGCCACGATGTCTCAAGAACTTCACCACCGGATTGATGCCGATGTGATCTTCCTGTCATAAACCATAGACGCGAGAC
~~~~~~~~~~~~~~~~~~~~~~~~~~~~~~~~~~~~~~~~~~~~~~~~~~~~~~~~~~~~~~~~~~~~~~~~~~~~~~~~~▶
CTGAAGCCAGTTACCTTCGGAAAAAGAGTTGGTAGCTCTTGATCCGGCAAACAAACCACCGCTGGTAGCGGTGGTTTTTTTGTTTGCAAGCAGCAGATTA
                                                                                                  3100.
GACTTCGGTCAATGGAAGCCTTTTTCTCAACCATCGAGAACTAGGCCGTTTGTTTGGTGGCGACCATCGCCACCAAAAAAACAAACGTTCGTCGTCTAAT

CGCGCAGAAAAAAAGGATCTCAAGAAGATCCTTTGATCTTTTCTACGGGGTCTGACGCTCAGTGGAACGAAAACTCACGTTAAGGGATTTTGGTCATGAG
                                                                                                  3200.
GCGCGTCTTTTTTTCCTAGAGTTCTTCTAGGAAACTAGAAAAGATGCCCCAGACTGCGAGTCACCTTGCTTTTGAGTGCAATTCCCTAAAACCAGTACTC

ATTATCAAAAAGGATCTTCACCTAGATCCTTTTAAATTAAAAATGAAGTTTTAAATCAATCTAAAGTATATATGAGTAAACTTGGTCTGACAGTTACCAA
                                                                                                  3300.
TAATAGTTTTTCCTAGAAGTGGATCTAGGAAAATTTAATTTTTACTTCAAAATTTAGTTAGATTTCATATATACTCATTTGAACCAGACTGTCAATGGTT

TGCTTAATCAGTGAGGCACCTATCTCAGCGATCTGTCTATTTCGTTCATCCATAGTTGCCTGACTCCCCGTCGTGTAGATAACTACGATACGGGAGGGCT
                                                                                                  3400.
ACGAATTAGTCACTCCGTGGATAGAGTCGCTAGACAGATAAAGCAAGTAGGTATCAACGGACTGAGGGGCAGCACATCTATTGATGCTATGCCCTCCCGA

TACCATCTGGCCCCAGTGCTGCAATGATACCGCGAGACCCACGCTCACCGGCTCCAGATTTATCAGCAATAAACCAGCCAGCCGGAAGGGCCGAGCGCAG
                                                                                                  3500.
ATGGTAGACCGGGGTCACGACGTTACTATGGCGCTCTGGGTGCGAGTGGCCGAGGTCTAAATAGTCGTTATTTGGTCGGTCGGCCTTCCCGGCTCGCGTC

AAGTGGTCCTGCAACTTTATCCGCCTCCATCCAGTCTATTAATTGTTGCCGGGAAGCTAGAGTAAGTAGTTCGCCAGTTAATAGTTTGCGCAACGTTGTT
                                                                                                  3600.
TTCACCAGGACGTTGAAATAGGCGGAGGTAGGTCAGATAATTAACAACGGCCCTTCGATCTCATTCATCAAGCGGTCAATTATCAAACGCGTTGCAACAA
        Pst I
GCCATTGCTGCAGGCATCGTGGTGTCACGCTCGTCGTTTGGTATGGCTTCATTCAGCTCCGGTTCCCAACGATCAAGGCGAGTTACATGATCCCCCATGT
                                                                                                  3700.
CGGTAACGACGTCCGTAGCACCACAGTGCGAGCAGCAAACCATACCGAAGTAAGTCGAGGCCAAGGGTTGCTAGTTCCGCTCAATGTACTAGGGGGTACA
                                           Pvu I
TGTGCAAAAAAGCGGTTAGCTCCTTCGGTCCTCCGATCGTTGTCAGAAGTAAGTTGGCCGCAGTGTTATCACTCATGGTTATGGCAGCACTGCATAATTC
                                                                                                  3800.
ACACGTTTTTTCGCCAATCGAGGAAGCCAGGAGGCTAGCAACAGTCTTCATTCAACCGGCGTCACAATAGTGAGTACCAATACCGTCGTGACGTATTAAG

TCTTACTGTCATGCCATCCGTAAGATGCTTTTCTGTGACTGGTGAGTACTCAACCAAGTCATTCTGAGAATAGTGTATGCGGCGACCGAGTTGCTCTTGC
                                                                                                  3900.
AGAATGACAGTACGGTAGGCATTCTACGAAAAGACACTGACCACTCATGAGTTGGTTCAGTAAGACTCTTATCACATACGCCGCTGGCTCAACGAGAACG

CCGGCGTCAACACGGGATAATACCGCGCCACATAGCAGAACTTTAAAAGTGCTCATCATTGGAAAACGTTCTTCGGGGCGAAAACTCTCAAGGATCTTAC
                                                                                                  4000.
GGCCGCAGTTGTGCCCTATTATGGCGCGGTGTATCGTCTTGAAATTTTCACGAGTAGTAACCTTTTGCAAGAAGCCCCGCTTTTGAGAGTTCCTAGAATG

CGCTGTTGAGATCCAGTTCGATGTAACCCACTCGTGCACCCAACTGATCTTCAGCATCTTTTACTTTCACCAGCGTTTCTGGGTGAGCAAAAACAGGAAG
                                                                                                  4100.
GCGACAACTCTAGGTCAAGCTACATTGGGTGAGCACGTGGGTTGACTAGAAGTCGTAGAAAATGAAAGTGGTCGCAAAGACCCACTCGTTTTTGTCCTTC

GCAAAATGCCGCAAAAAAGGGAATAAGGGCGACACGGAAATGTTGAATACTCATACTCTTCCTTTTTCAATATTATTGAAGCATTTATCAGGGTTATTGT
                                                                                                  4200.
CGTTTTACGGCGTTTTTTCCCTTATTCCCGCTGTGCCTTTACAACTTATGAGTATGAGAAGGAAAAAGTTATAATAACTTCGTAAATAGTCCCAATAACA
                                                                                 ◀~~~~~Apr promoter
CTCATGAGCGGATACATATTTGAATGTATTTAGAAAAATAAACAAATAGGGGTTCCGCGCACATTTCCCCGAAAAGTGCCACCTGACGTCTAAGAAACCA
                                                                                                  4300.
GAGTACTCGCCTATGTATAAACTTACATAAATCTTTTTATTTGTTTATCCCCAAGGCGCGTGTAAAGGGGCTTTTCACGGTGGACTGCAGATTCTTTGGT
                                                        Eco RI
TTATTATCATGACATTAACCTATAAAAATAGGCGTATCACGAGGCCCTTTCGTCTTCAAGAA(TTC)
AATAATAGTACTGTAATTGGATATTTTTATCCGCATAGTGCTCCGGGAAAGCAGAAGTTCTT   pBR322
```

APPENDIX 3

RESTRICTION MAPS
AND NUCLEOTIDE SEQUENCES

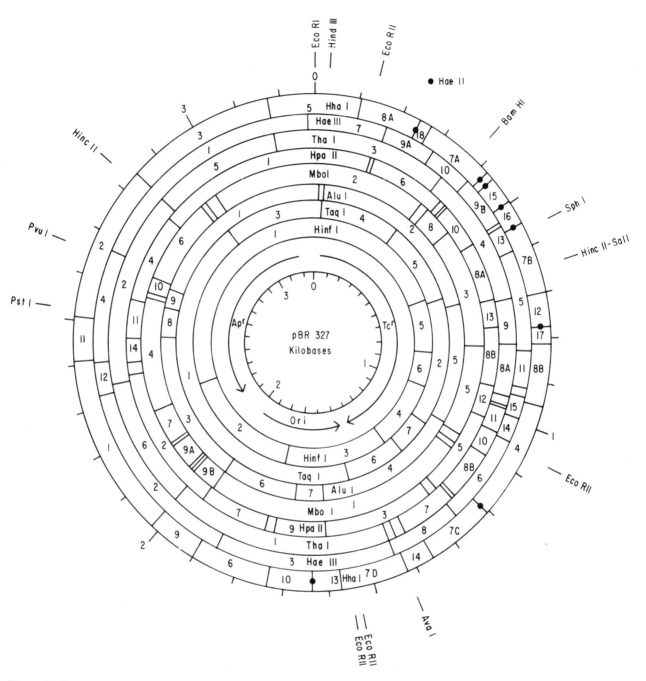

Figure A3.3
Restriction endonuclease cleavage map of pBR327. (From Soberon et al., Gene 9:287, 1980.)

207

APPENDIX 3

RESTRICTION MAPS AND NUCLEOTIDE SEQUENCES

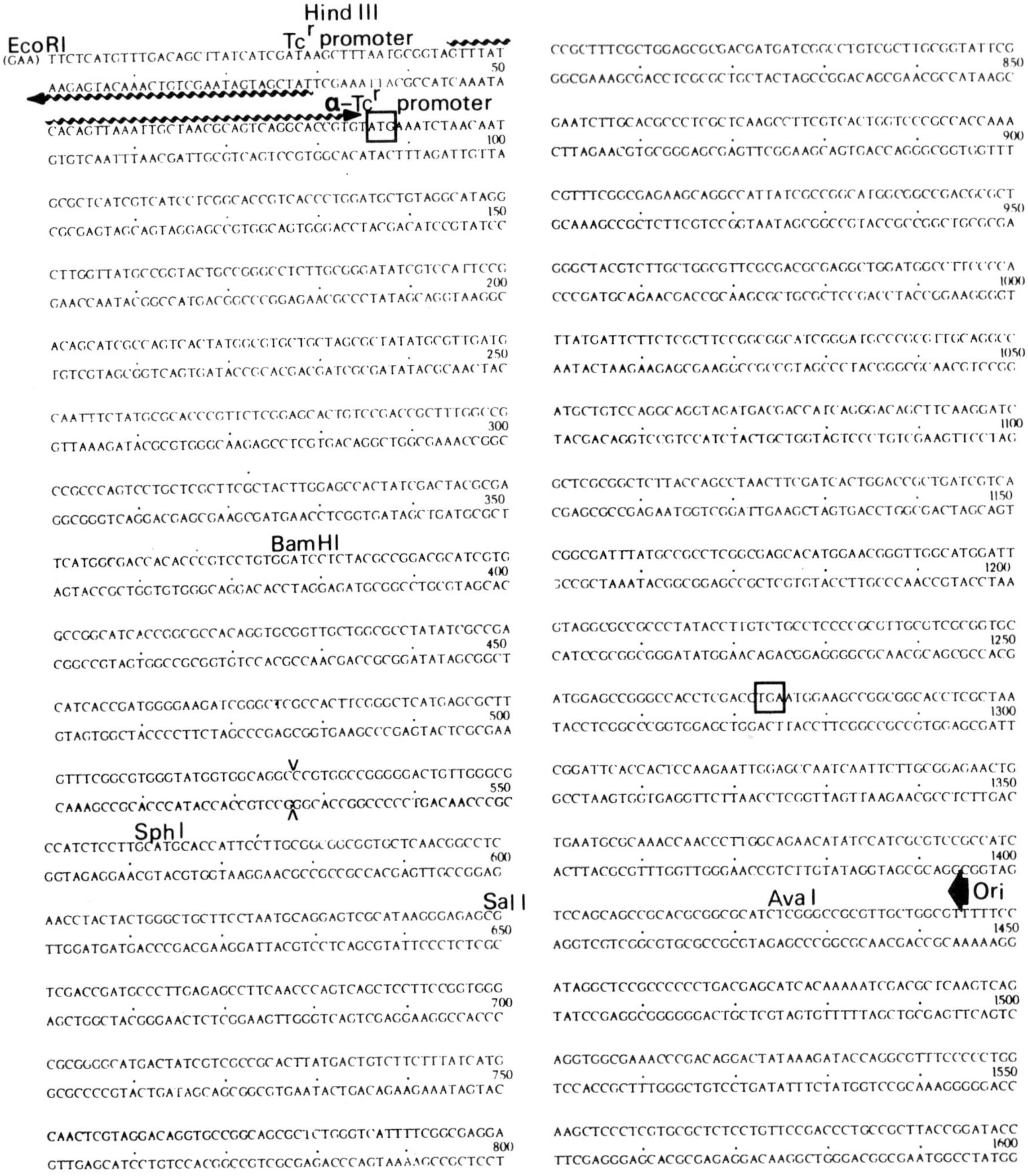

Figure A3.4
Nucleotide sequence of pBR327. (From Soberon et al., Gene 9: 287, 1980.)

APPENDIX 3

RESTRICTION MAPS AND NUCLEOTIDE SEQUENCES

```
TGTCCGCCTTTCTCCCTTCGGGAAGCGTGGCGCTTTCTCAATGCTCACGC
ACAGGCGGAAAGAGGGAAGCCCTTCGCACCGCGAAAGAGTTACGAGTGCG
                                                1650

TGTAGGTATCTCAGTTCGGTGTAGGTCGTTCGCTCCAAGCTGGGCTGTGT
ACATCCATAGAGTCAAGCCACATCCAGCAAGCGAGGTTCGACCCGACACA
                                                1700

GCACGAACCCCCCGTTCAGCCCGACCGCTGCGCCTTATCCGGTAACTATC
CGTGCTTGGGGGGCAAGTCGGGCTGGCGACGCGGAATAGGCCATTGATAG
                                                1750

GTCTTGAGTCCAACCCGGTAAGACACGACTTATCGCCACTGGCAGCAGCC
CAGAACTCAGGTTGGGCCATTCTGTGCTGAATAGCGGTGACCGTCGTCGG
                                                1800

ACTGGTAACAGGATTAGCAGAGCGAGGTATGTAGGCGGTGCTACAGAGTT
TGACCATTGTCCTAATCGTCTCGCTCCATACATCCGCCACGATGTCTCAA
                                                1850
                           104 base RNA ∼∼∼∼∼∼∼
CTTGAAGTGGTGGCCTAACTACGGCTACACTAGAAGGACAGTATTTGGTA
GAACTTCACCACCGGATTGATGCCGATGTGATCTTCCTGTCATAAACCAT
                                                1900

TCTGCGCTCTGCTGAAGCCAGTTACCTTCGGAAAAAGAGTTGGTAGCTCT
AGACGCGAGACGACTTCGGTCAATGGAAGCCTTTTTCTCAACCATCGAGA
                                                1950
∼∼∼∼∼∼∼∼∼∼∼∼∼∼∼∼∼∼∼∼∼∼∼∼∼∼∼∼∼∼∼∼∼∼∼∼∼∼∼∼∼∼∼→
TGATCCGGCAAACAAACCACCGCTGGTAGCGGTGGTTTTTTTGTTTGCAA
ACTAGGCCGTTTGTTTGGTGGCGACCATCGCCACCAAAAAAACAAACGTT
                                                2000

GCAGCAGATTACGCGCAGAAAAAAAGGATCTCAAGAAGATCCTTTGATCT
CGTCGTCTAATGCGCGTCTTTTTTTCCTAGAGTTCTTCTAGGAAACTAGA
                                                2050

TTTCTACGGGGTCTGACGCTCAGTGGAACGAAAACTCACGTTAAGGGATT
AAAGATGCCCCAGACTGCGAGTCACCTTGCTTTTGAGTGCAATTCCCTAA
                                                2100

TTGGTCATGAGATTATCAAAAAGGATCTTCACCTAGATCCTTTTAAATTA
AACCAGTACTCTAATAGTTTTTCCTAGAAGTGGATCTAGGAAAATTTAAT
                                                2150

AAAATGAAGTTTTAAATCAATCTAAAGTATATATGAGTAAACTTGGTCTG
TTTTACTTCAAAATTTAGTTAGATTTCATATATACTCATTTGAACCAGAC
                                                2200

ACAGTTACCAATGCTTAATCAGTGAGGCACCTATCTCAGCGATCTGTCTA
TGTCAATGGTTACGAATTAGTCACTCCGTGGATAGAGTCGCTAGACAGAT
                                                2250

TTTCGTTCATCCATAGTTGCCTGACTCCCCGTCGTGTAGATAACTACGAT
AAAGCAAGTAGGTATCAACGGACTGAGGGGCAGCACATCTATTGATGCTA
                                                2300

ACGGGAGGGCTTACCATCTGGCCCCAGTGCTGCAATGATACCGCGAGACC
TGCCCTCCCGAATGGTAGACCGGGGTCACGACGTTACTATGGCGCTCTGG
                                                2350

CACGCTCACCGGCTCCAGATTTATCAGCAATAAACCAGCCAGCCGGAAGG
GTGCGAGTGGCCGAGGTCTAAATAGTCGTTATTTGGTCGGTCGGCCTTCC
                                                2400

GCCGAGCGCAGAAGTGGTCCTGCAACTTTATCCGCCTCCATCCAGTCTAT
CGGCTCGCGTCTTCACCAGGACGTTGAAATAGGCGGAGGTAGGTCAGATA
                                                2450
```

```
TAATTGTTGCCGGGAAGCTAGAGTAAGTAGTTCGCCAGTTAATAGTTTGC
ATTAACAACGGCCCTTCGATCTCATTCATCAAGCGGTCAATTATCAAACG
                                                2500
              PstI
GCAACGTTGTTGCCATTGCTGCAGGCATCGTGGTGTCACGCTCGTCGTTT
CGTTGCAACAACGGTAACGACGTCCGTAGCACCACAGTGCGAGCAGCAAA
                                                2550

GGTATGGCTTCATTCAGCTCCGGTTCCCAACGATCAAGGCGAGTTACATG
CCATACCGAAGTAAGTCGAGGCCAAGGGTTGCTAGTTCCGCTCAATGTAC
                                                2600
                                          PvuI
ATCCCCCATGTTGTGCAAAAAAGCGGTTAGCTCCTTCGGTCCTCCGATCG
TAGGGGGTACAACACGTTTTTTCGCCAATCGAGGAAGCCAGGAGGCTAGC
                                                2650

TTGTCAGAAGTAAGTTGGCCGCAGTGTTATCACTCATGGTTATGGCAGCA
AACAGTCTTCATTCAACCGGCGTCACAATAGTGAGTACCAATACCGTCGT
                                                2700

CTGCATAATTCTCTTACTGTCATGCCATCCGTAAGATGCTTTTCTGTGAC
GACGTATTAAGAGAATGACAGTACGGTAGGCATTCTACGAAAAGACACTG
                                                2750

TGGTGAGTACTCAACCAAGTCATTCTGAGAATAGTGTATGCGGCGACCGA
ACCACTCATGAGTTGGTTCAGTAAGACTCTTATCACATACGCCGCTGGCT
                                                2800

GTTGCTCTTGCCCGGCCGTCAACACGGGATAATACCGCGCCACATAGCAGA
CAACGAGAACGGGCCGCAGTTGTGCCCTATTATGGCGCGGTGTATCGTCT
                                                2850

ACTTTAAAAGTGCTCATCATTGGAAAACGTTCTTCGGGGCGAAAACTCTC
TGAAATTTTCACGAGTAGTAACCTTTTGCAAGAAGCCCCGCTTTTGAGAG
                                                2900

AAGGATCTTACCGCTGTTGAGATCCAGTTCGATGTAACCCACTCGTGCAC
TTCCTAGAATGGCGACAACTCTAGGTCAAGCTACATTGGGTGAGCACGTG
                                                2950

CCAACTGATCTTCAGCATCTTTTACTTTCACCAGCGTTTCTGGGTGAGCA
GGTTGACTAGAAGTCGTAGAAAATGAAAGTGGTCGCAAAGACCCACTCGT
                                                3000

AAAACAGGAAGGCAAAATGCCGCAAAAAAGGGAATAAGGGCGACACGGAA
TTTTGTCCTTCCGTTTTACGGCGTTTTTTCCCTTATTCCCGCTGTGCCTT
                                                3050

ATGTTGAATACTCATACTCTTCCTTTTTCAATATTATTGAAGCATTTATC
TACAACTTATGAGTATGAGAAGGAAAAAGTTATAATAACTTCGTAAATAG
                                                3100
                    ←∼∼∼∼∼∼∼∼∼∼∼∼∼∼∼∼∼∼∼∼
AGGGTTATTGTCTCATGAGCGGATACATATTTGAATGTATTTAGAAAAAT
TCCCAATAACAGAGTACTCGCCTATGTATAAACTTACATAAATCTTTTTA
                                                3150

AAACAAATAGGGGTTCCGCGCACATTTCCCCGAAAAGTGCCACCTGACGT
TTTGTTTATCCCCAAGGCGCGTGTAAAGGGGCTTTTCACGGTGGACTGCA
                                                3200

CTAAGAAACCATTATTATCATGACATTAACCTATAAAAATAGGCGTATCA
GATTCTTTGGTAATAATAGTACTGTAATTGGATATTTTTATCCGCATAGT
                                                3250

CGAGGCCCTTTCGTCTTCAAGAA (TTC)
GCTCCGGGAAAGCAGAAGTTCTT                    pBR327
```

209

APPENDIX 3

RESTRICTION MAPS AND NUCLEOTIDE SEQUENCES

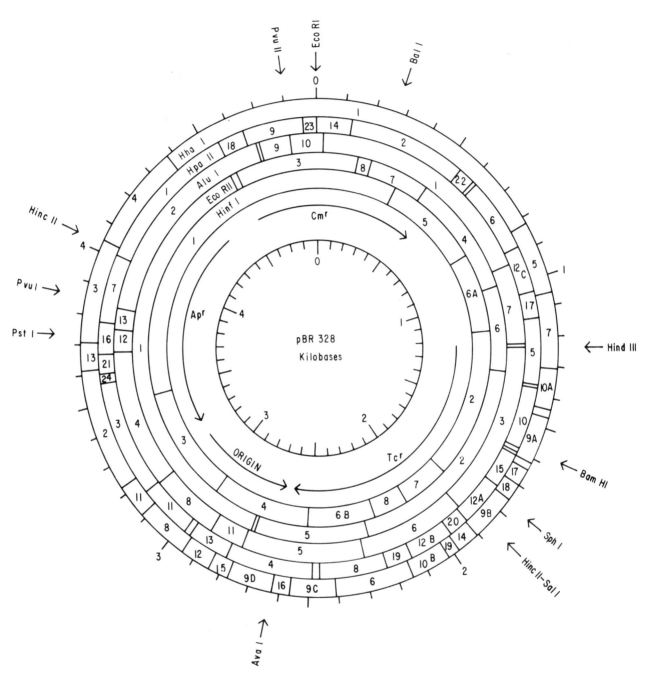

Figure A3.5
Restriction endonuclease cleavage map of pBR328. (From Soberon et al., Gene 9:287, 1980.)

APPENDIX 3

RESTRICTION MAPS
AND NUCLEOTIDE SEQUENCES

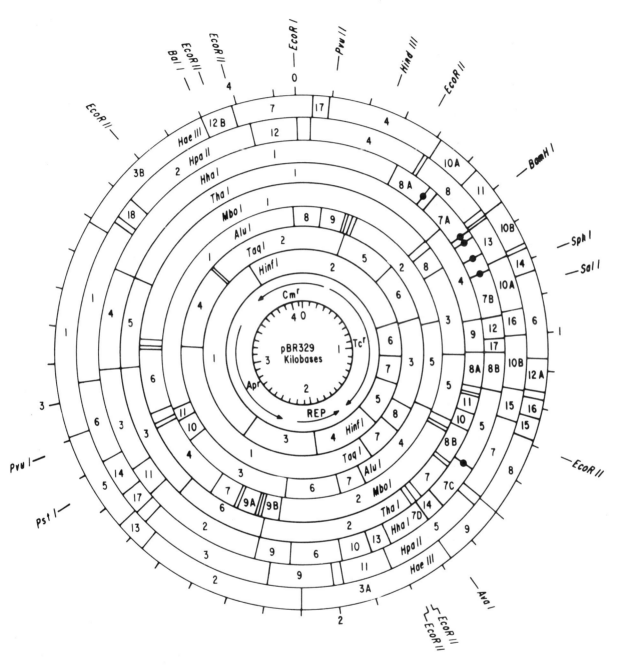

Figure A3.6
Restriction endonuclease cleavage map of pBR329. (From Covarrubias and Bolivar, Gene 17:79, 1982.)

Eco RI
(GAA) TTCCGGATGAGCATTCATCAGGCGGGCAAGAATGTGAATAAAGGCCGGAT   50
      AAGGCCTACTCGTAAGTAGTCCGCCCGTTCTTACACTTATTTCCGGCCTA

      AAAACTTGTGCTTATTTTTCTTTACGGTCTTTAAAAAGGCCGTAATATCC   100
      TTTTGAACACGAATAAAAAGAAATGCCAGAAATTTTTCCGGCATTATAGG
Pvu II
      AGCTGAACGGTCTGGTTATAGGTACATTGAGCAACTGACTGAAATGCCTC   150
      TCGACTTGCCAGACCAATATCCATGTAACTCGTTGACTGACTTTACGGAG

      AAAATGTTCTTTACGATGCCATTGGGATATATCAACGGTGGTATATCCAG   200
      TTTTACAAGAAATGCTACGGTAACCCTATATAGTTGCCACCATATAGGTC

      TGATTTTTTTCTCCATTTTAGCTTCCTTAGCTCCTGATGTTTGACAGCTT   250
      ACTAAAAAAAGAGGTAAAATCGAAGGAATCGAGGACTACAAACTGTCGAA
                    Cm^r
            Hind III
      ATCATCGATAAGCTTTAATGCGGTAGTTTATCACAGTTAAATTGCTAACG   300
      TAGTAGCTATTCGAAATTACGCCATCAAATAGTGTCAATTTAACGATTGC
            α-Tc^r
      CAGTCAGGCACCGTGTATGAAATCTAACAATGCGCTCATCGTCATCCTCG   350
      GTCAGTCCGTGGCACATACTTTAGATTGTTACGCGAGTAGCAGTAGGAGC
                    Tc^r
      GCACCGTCACCCTGGATGCTGTAGGCATAGGCTTGGTTATGCCGGTACTG   400
      CGTGGCAGTGGGACCTACGACATCCGTATCCGAACCAATACGGCCATGAC

      CCGGGCCTCTTGCGGGATATCGTCCATTCCGACAGCATCGCCAGTCACTA   450
      GGCCCGGAGAACGCCCTATAGCAGGTAAGGCTGTCGTAGCGGTCAGTGAT

      TGGCGTGCTGCTAGCGCTATATGCGTTGATGCAATTTCTATGCGCACCCG   500
      ACCGCACGACGATCGCGATATACGCAACTACGTTAAAGATACGCGTGGGC

      TTCTCGGAGCACTGTCCGACCGCTTTGGCCGCCGCCCAGTCCTGCTGCT   550
      AAGAGCCTCGTGACAGGCTGGCGAAACCGGCGGCGGGTCAGGACGAGCGA

      TCGCTACTTGGAGCCACTATCGACTACGCGATCATGGCGACCACACCCGT   600
      AGCGATGAACCTCGGTGATAGCTGATGCGCTAGTACCGCTGGTGTGGGCA
      BamHI
      CCTGTGGATCCTCTACGCCGGACGCATCGTGGCCGGCATCACCGGCGCCA   650
      GGACACCTAGGAGATGCGGCCTGCGTAGCACCGGCCGTAGTGGCCGCGGT

      CAGGTGCGGTTGCTGGCGCCTATATCGCCGACATCACCGATGGGGAAGAT   700
      GTCCACGCCAACGACCGCGGATATAGCGGCTGTAGTGGCTACCCCTTCTA

      CGGGCTCGCCACTTCGGGCTCATGAGCGCTTGTTTCGGCGTGGGTATGGT   750
      GCCCGAGCGGTGAAGCCCGAGTACTCGCGAACAAAGCCGCACCCATACCA
          v   **Sph I**
      GGCAGGCCCGTGGCCGGGGACTGTTGGGCGCCATCTCCTTGCATGCACC   800
      CCGTCCGGGCACCGGCCCCCTGACAACCCGCGGTAGAGGAACGTAGCTGG
          λ
      ATTCCTTGCGGCGGCGGTGCTCAACGGCCTCAACCTACTACTGGGCTGCT   850
      TAAGGAACGCCGCCGCCACGAGTTGCCGGAGTTGGATGATGACCCGACGA
                              Sal I
      TCCTAATGCAGGAGTCGCATAAGGGAGAGCGTCGACCGATGCCCTTGAGA   900
      AGGATTACGTCCTCAGCGTATTCCCTCTCGCAGCTGGCTACGGGAACTCT

      GCCTTCAACCCAGTCAGCTCCTTCCGGTGGGCGCGGGGCATGACTATCGT   950
      CGGAAGTTGGGTCAGTCGAGGAAGGCCACCCGCGCCCCGTACTGATAGCA

      CGCCGCACTTATGACTGTCTTCTTTATCATGCAACTCGTAGGACAGGTGC   1000
      GCGGCGTGAATACTGACAGAAGAAATAGTACGTTGAGCATCCTGTCCACG

      CGGCAGCGCTCTGGGTCATTTTCGGCGAGGACCGCTTTCGCTGGAGCGCG   1050
      GCCGTCGCGAGACCCAGTAAAAGCCGCTCCTGGCGAAAGCGACCTCGCGC

      ACGATGATCGGCCTGTCGCTTGCGGTATTCGGAATCTTGCACGCCCTCGC   1100
      TGCTACTAGCCGGACAGCGAACGCCATAAGCCTTAGAACGTGCGGGAGCG

      TCAAGCCTTCGTCACTGGTCCCGCCACCAAACGTTTCGGCGAGAAGCAGG   1150
      AGTTCGGAAGCAGTGACCAGGGCGGTGGTTTGCAAAGCCGCTCTTCGTCC

      CCATTATCGCCGGCATGGCGGCCGACGCGCTGGGCTACGTCTTGCTGGCG   1200
      GGTAATAGCGGCCGTACCGCCGGCTGCGCGACCCGATGCAGAACGACCGC

      TTCGCGACGCGAGGCTGGATGGCCTTCCCCATTATGATTCTTCTCGCTTC   1250
      AAGCGCTGCGCTCCGACCTACCGGAAGGGGTAATACTAAGAAGAGCGAAG

      CGGCGGCATCGGGATGCCCGCGTTGCAGGCCATGCTGTCCAGGCAGGTAG   1300
      GCCGCCGTAGCCCTACGGGCGCAACGTCCGGTACGACAGGTCCGTCCATC

      ATGACGACCATCAGGGACAGCTTCAAGGATCGCTCGCGGCTCTTACCAGC   1350
      TACTGCTGGTAGTCCCTGTCGAAGTTCCTAGCGAGCGCCGAGAATGGTCG

      CTAACTTCGATCACTGGACCGCTGATCGTCACGGCGATTTATGCCGCCTC   1400
      GATTGAAGCTAGTGACCTGGCGACTAGCAGTGCCGCTAAATACGGCGGAG

      GGCGAGCACATGGAACGGGTTGGCATGGATTGTAGGCGCCGCCCTATACC   1450
      CCGCTCGTGTACCTTGCCCAACCGTACCTAACATCCGCGGCGGGATATGG

      TTGTCTGCCTCCCCGCGTTGCGTCGCGGTGCATGGAGCCGGGCCACCTCG   1500
      AACAGACGGAGGGGCGCAACGCAGCGCCACGTACCTCGGCCCGGTGGAGC

      ACCTGAATGGAAGCCGGCGGCACCTCGCTAACGGATTCACCACTCCAAGA   1550
      TGGACTTACCTTCGGCCGCCGTGGAGCGATTGCCTAAGTGGTGAGGTTCT

      ATTGGAGCCAATCAATTCTTGCGGAGAACTGTGAATGCGCAAACCAACCC   1600
      TAACCTCGGTTAGTTAAGAACGCCTCTTGACACTTACGCGTTTGGTTGGG

      TTGGCAGAACATATCCATCGCGTCCGCCATCTCCAGCAGCCGCACGCGGC   1650
      AACGTCTTGTATAGGTAGCGCAGGCGGTAGAGGTCGTCGGCGTGCGCCG
                                  ← **ori**
      GCATCTCGGGCCGCGTTGCTGGCGTTTTTCCATAGGCTCCGCCCCCCTGA   1700
      CGTAGAGCCCGGCGCAACGACCGCAAAAAGGTATCCGAGGCGGGGGGACT

      CGAGCATCACAAAAATCGACGCTCAAGTCAGAGGTGGCGAAACCCGACAG   1750
      GCTCGTAGTGTTTTTAGCTGCGAGTTCAGTCGTTCACCGCTTTGGGCTGTC

      GACTATAAAGATACCAGGCGTTTCCCCCTGGAAGCTCCCTCGTGCGCTCT   1800
      CTGATATTTCTATGGTCCGCAAAGGGGGACCTTCGAGGGAGCACGCGAGA

      CCTGTTCCGACCCTGCCGCTTACCGGATACCTGTCCGCCTTTCTCCCTTC   1850
      GGACAAGGCTGGGACGGCGAATGGCCTATGGACAGGCGGAAAGAGGGAAG

      GGGAAGCGTGGCGCTTTCTCAATGCTCACGCTGTAGGTATCTCAGTTCGG   1900
      CCCTTCGCACCGCGAAAGAGTTACGAGTGCGACATCCATAGACTCAAGCC

      TGTAGGTCGTTCGCTCCAAGCTGGGCTGTGTGCACGAACCCCCCGTTCAG   1950
      ACATCCAGCAAGCGAGGTTCGACCCGACACACGTGCTTGGGGGGCAAGTC

      CCCGACCGCTGCGCCTTATCCGGTAACTATCGTCTTGAGTCCAACCCGGT   2000
      GGGCTGGCGACGCGGAATAGGCCATTGATAGCAGAACTCAGGTTGGGCCA

      AAGACACGACTTATCGCCACTGGCAGCAGCCACTGGTAACAGGATTAGCA   2050
      TTCTGTGCTGAATAGCGGTGACCGTCGTCGGTGACCATTGTCCTAATCGT

      GAGCGAGGTATGTAGGCGGTGCTACAGAGTTCTTGAAGTGGTGGCCTAAC   2100
      CTCGCTCCATACATCCGCCACGATGTCTTAAGAACTCCACCACCGGATTG

      104 base RNA ~~~~~~~
      TACGGCTACACTAGAAGGACAGTATTTGGTATCTGCGCTCTGCTGAAGCC   2150
      ATGCCGATGTGATCTTCCAGTCATAAACCATAGACGCGATACGACTTCGG

Figure A3.7

Nucleotide sequence of pBR329. (From Covarrubias and Bolivar, *Gene* 17:79, 1982.)

```
AGTTACCTTCGGAAAAAGAGTTGGTAGCTCTTGATCCGGCAAACAAACCA
TCAATGGAAGCCTTTTTCTCAACCATCGAGAACTAGGCCGTTTGTTTGGT        2200

CCGCTGGTAGCGGTGGTTTTTTTGTTTGCAAGCAGCAGATTACGCGCAGA
GGCGACCATCGCCACCAAAAAAACAAACGTTCGTCGTCTAATGCGCGTCT        2250

AAAAAAGGATCTCAAGAAGATCCTTTGATCTTTTCTACGGGGTCTGACGC
TTTTTTCCTAGAGTTCTTCTAGGAAACTAGAAAAGATGCCCCAGACTGCG        2300

TCAGTGGAACGAAAACTCACGTTAAGGGATTTTGGTCATGAGATTATCAA
AGTCACCTTGCTTTTGAGTGCAATTCCCTAAAACCAGTACTCTAATAGTT        2350

AAAGGATCTTCACCTAGATCCTTTTAAATTAAAAATGAAGTTTTAAATCA
TTTCCTAGAAGTGGATCTAGGAAAATTTAATTTTTACTTCAAAATTTAGT        2400

ATCTAAAGTATATATGAGTAAACTTGGTCTGACAGTTACCAATGCTTAAT
TAGATTTCATATATACTCATTTGAACCAGACTGTCAATGGTTACGAATTA       2450

CAGTGAGGCACCTATCTCAGCGATCTGTCTATTTCGTTCATCCATAGTTG
GTCACTCCGTGGATAGAGTCGCTAGACAGATAAAGCAAGTAGGTATCAAC        2500

CCTGACTCCCCGTCGTGTAGATAACTACGATACGGGAGGGCTTACCATCT
GGACTGAGGGGCAGCACATCTATTGATGCTATGCCCTCCCGAATGGTAGA        2550

GGCCCCAGTGCTGCAATGATACCGCGAGACCCACGCTCACCGGCTCCAGA
CCGGGGTCACGACGTTACTATGGCGCTCTGGGTGCGAGTGGCCGAGGTCT        2600

TTTATCAGCAATAAACCAGCCAGCCGGAAGGGCCGAGCGCAGAAGTGGTC
AAATAGTCGTTATTTGGTCGGTCGGCCTTCCCGGCTCGCGTCTTCACCAG        2650

CTGCAACTTTATCCGCCTCCATCCAGTCTATTAATTGTTGCCGGGAAGCT
GACGTTGAAATAGGCGGAGGTAGGTCAGATAATTAACAACGGCCCTTCGA        2700

    AGAGTAAGTAGTTCGCCAGTTAATAGTTTGCGCAACGTTGTTGCCATTGC
    TCTCATTCATCAAGCCGGTCAATTATCAAACGCGTTGCAACAACGGTAACG    2750
Pst I
    TGCAGGCATCGTGGTGTCACGCTCGTCGTTTGGTATGGCTTCATTCAGCT
    ACGTCCGTAGCACCACAGTGCGAGCAGCAAACCATACCGTAGTAAGTCGA    2800

    CCGGTTCCCAACGATCAAGGCGAGTTACATGATCCCCCATGTTGTGCAAA
    GGCCAAGGGTTGCTAGTTCCGCTCAATGTACTAGGGGGTACAACACGTTT    2850
                        Pvu I
    AAAGCGGTTAGCTCCTTCGGTCCTCCGATCGTTGTCAGAAGTAAGTTGGC
    TTTCGCCAATCGAGGAAGCCAGGAGGCTAGCAACAGTCTTCATTCAACCG    2900

    CGCAGTGTTATCACTCATGGTTATGGCAGCACTGCATAATTCTCTTACTG
    GCGTCACAATAGTGAGTACCAATACCGTCGTGACGTATTAAGAGAATGAC    2950

    TCATGCCATCCGTAAGATGCTTTTCTGTGACTGGTGAGTACTCAACCAAG
    AGTACGGTAGGCATTCTACGAAAAGACACTGACCACTCATGAGTTGGTTC    3000

    TCATTCTGAGAATAGTGTATGCGGCGACCGAGTTGCTCTTGCCCGGCGTC
    AGTAAGACTCTTATCACATACGCCGCTGGCTCAACGAGAACGGGCCGCAG    3050

    AACACGGGATAATACCGCGCCACATAGCAGAACTTTAAAAGTGCTCATCA
    TTGTGCCCTATTATGGCGCGGTGTATCGTCTTGAAATTTTCACGAGGAGT    3100

    TTGGAAAACGTTCTTCGGGGCGAAAACTCTCAAGGATCTTACCGCTGTTG
    AACCTTTTGCAAGAAGCCCCGCTTTTGAGAGTTCCTAGAATGGCGACAAC    3150

    AGATCCAGTTCGATGTAACCCACTCGTGCACCCAACTGATCTTCAGCATC
    TCTAGGTCAAGCTACATTGGGTGAGCACGTGGGTTGACTAGAAGTCGTAG    3200

    TTTTACTTTCACCAGCGTTTCTGGGTGAGCAAAAACAGGAAGGCAAAATG
    AAAATGAAAGTGGTCGCAAAGACCCACTCGTTTTTGTCCTTCCGTTTTAC    3250

CCGCAAAAAAGGGAATAAGGGCGACACGGAAATGTTGAATACTCATACTC                           Ap^r
GGCGTTTTTTCCCTTATTCCCGCTGTGCCTTTACAACTTATGAGTATGAG        3300

TTCCTTTTTCAATATTATTTAAGCATTTATCAGGGTTATTGTCTCATGAG
AAGGAAAAAGTTATAATAACTTCGTAAATAGTCCCAATAACAGAGTACTC        3350

CGGATACATATTTGAATGTATTTAGAAAAATAAACAAATAGGGGTTCCGC
GCCTATGTATAAACTTACATAAATCTTTTTATTTGTTTATCCCCAAGGCG        3400

GCACATTTCCCCGAAAAGTGCCACCTGACGTCTAAGAAACCATTATTATC
CGTGTAAAGGGGCTTTTCACGGTGGACTGCAGATTCTTTGGTAATAATAG        3450

ATGACATTAACCTATAAAAATAGGCGTATCACGAGGCCCTTTCGTCTAGG
TACTGTAATTGGATATTTTTATCCGCATAGTGCTCCGGGAAAGCAGATCC        3500

CATTTGAGAAGCACACGGTCACACTGCTTCCGGTAGTCAATAAACCGGTA
GTAAACTCTTCGTGTGCCAGTGTGACGAAGGCCATCAGTTATTTGGCCAT        3550

AACCAGCAATAGACATAAGCGGCTATTTAACGACCCTGCCCTGAACCGAC
TTGGTCGTTATCTGTATTCGCCGATAAATTGCTGGGACGGGACTTGGCTG        3600

GACCGGGTCGAATTTGCTTTCGAATTTCTGCCATTCATCCGCTTATTATC
CTGGCCCAGCTTAAACGAAAGCTTAAAGACGGTAAGTAGGCGAATAATAG        3650

ACTTATTCAGGCGTAGCACCAGGCGTTTAAGGGCACCAATAACTGCCTTA
TGAATAAGTCCGCATCGTGGTCCGCAAATTCCCGAGGTTATTGACGGAAT        3700

AAAAAATTACGCCCCGCCCTGCCACTCATCGCAGTACTGTTGTAATTCAT
TTTTTTAATGCGGGGCGGGACGGTGAGTAGCGTCATGACAACATTAAGTA        3750

TAAGCATTCTGCCGACATGGAAGCCATCACAAACGGCATGATGAACCTGA
ATTCGTAAGACGGCTGTACCTTCGGTAGTGTTTGCCGTACTACTTGGACT        3800

ATCGCCAGCGGCATCGACACCTTGTCGCCTTGCGTATAATATTTGCCCAT
TAGCGGTCGCCGTAGTCGTGGAACAGCGGAACGCATATTATAAACGGGTA        3850
                              Bal I
GGTGAAAACGGGGGCGAAGAAGTTGTCCATATTGGCCACGTTTAAATCAA
CCACTTTTGCCCCCGCTTCTTCAACAGGTATAACCGGTGCAAATTTAGTT        3900

AACTGGTGAAACTCACCCAGGGATTGGCTGAGACGAAAAACATATTCTCA
TTGACCACTTTGAGTGGGTCCCTAACCGACTCTGCTTTTTGTATAAGAGT        3950

ATAAACCCTTTAGGGAAATAGGCCAGGTTTTCACCGTAACACGCCACATC
TATTTGGGAAATCCCTTTATCCGGTCCAAAAGTGGCATTGTGCGGTGTAG        4000

TTGCGAATATATGTGTAGAAACTGCCGGAAATCGTCGTGGTATTCACTCC
AACGCTTATATACACATCTTTGACGGCCTTTAGCAGCACCATAAGTGAGG        4050

AGAGCGATGAAAACGTTTCAGTTTGCTCATGGAAAACGGTGTAACAAGGG
TCTCGCTACTTTTGCAAAGTCAAACGAGTACCTTTTGCCACATTGTTCCC        4100

TGAACACTATCCCTATATCACCAGCTCACCGTCTTTCATTGCCATACGAA
ACTTGTGATAGGGTATAGTGGTCGAGTGGCAGAAAGTAACGGTATGCCTT       (TTC)
                                                         EcoRI
pBR329
```

213

APPENDIX 3

RESTRICTION MAPS AND NUCLEOTIDE SEQUENCES

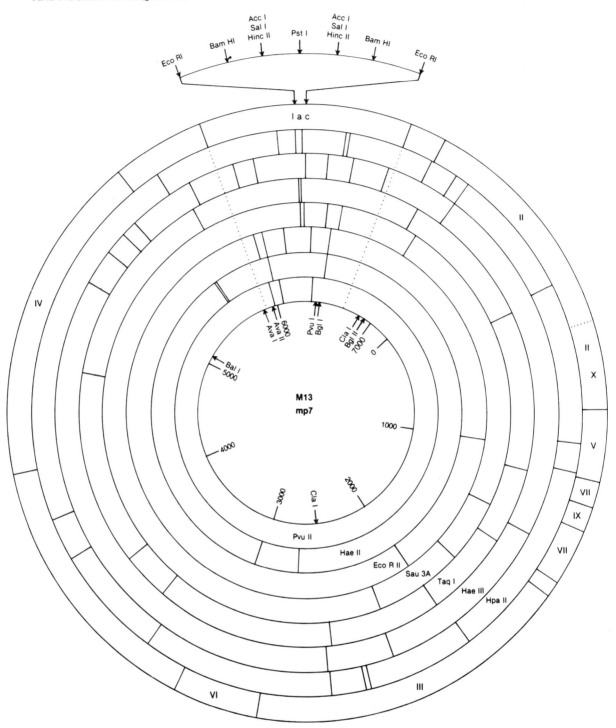

Figure A3.8
Restriction endonuclease cleavage map of M13mp7. (Prepared by Dr. Robert Blakesley (Bethesda Research Laboratories) from the primary sequence data of M13 (Van Wezenbeck et al., 1980, Gene 11:129) as modified by J. Messing. Copyright, Bethesda Research Laboratories, Inc.)

APPENDIX 3

RESTRICTION MAPS AND NUCLEOTIDE SEQUENCES

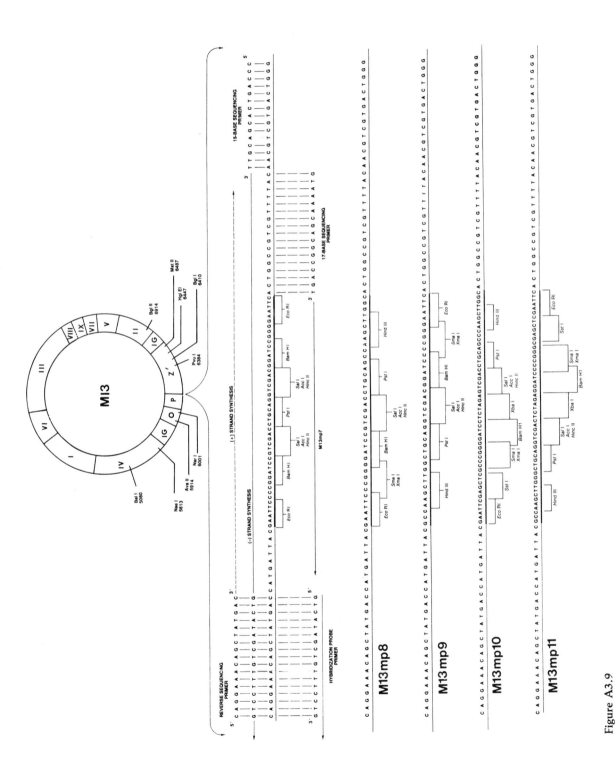

Figure A3.9
Nucleotide sequences of the multiple cloning sites in M13mp7, mp8, mp9, mp10, and mp11. (P-L Biochemicals, 1983©.)

APPENDIX 3

RESTRICTION MAPS
AND NUCLEOTIDE SEQUENCES

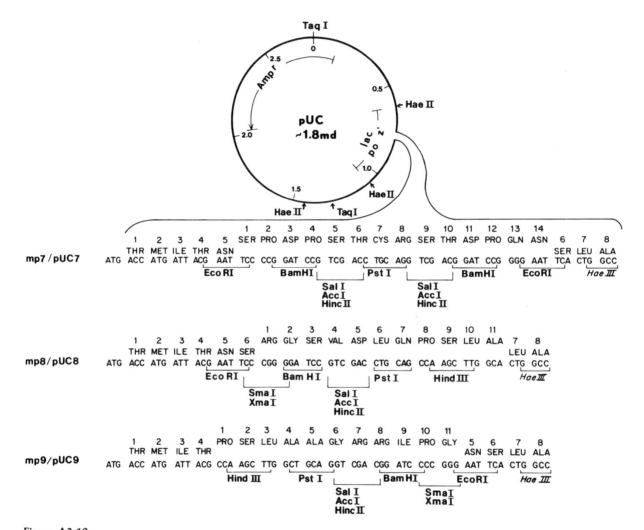

Figure A3.10
Nucleotide sequences of the multiple cloning sites in the plasmids, pUC7, pUC8 and pUC9. (From Vieira and Messing, Gene 19:259, 1982.)

APPENDIX 3

RESTRICTION MAPS
AND NUCLEOTIDE SEQUENCES

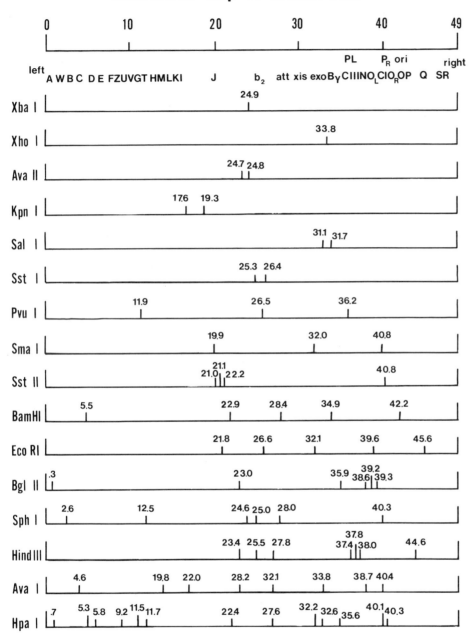

Figure A3.11
Restriction endonuclease cleavage map of bacteriophage lambda. Numbers represent kilobase coordinates from the left end of the lambda genome. (Modified from Bethesda Research Laboratories, 1981©.)

APPENDIX 3

RESTRICTION MAPS
AND NUCLEOTIDE SEQUENCES

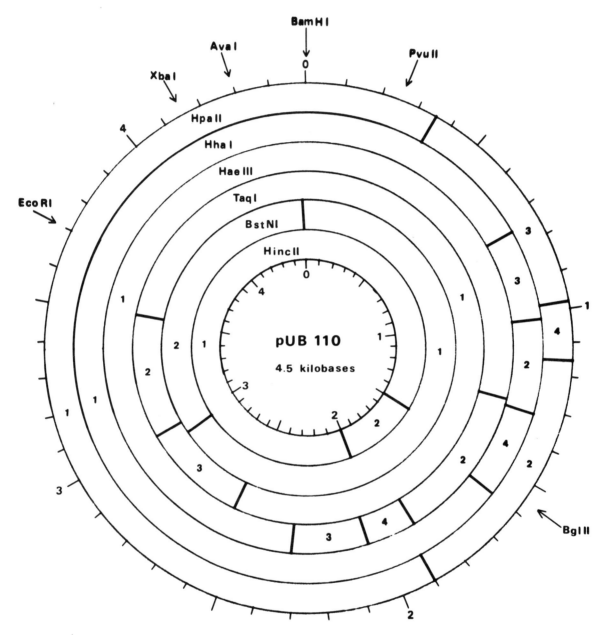

Figure A3.12
Restriction endonuclease cleavage map of pUB110. (From Jalanko and Palva, Gene 14:325, 1981.)

APPENDIX 3

RESTRICTION MAPS
AND NUCLEOTIDE SEQUENCES

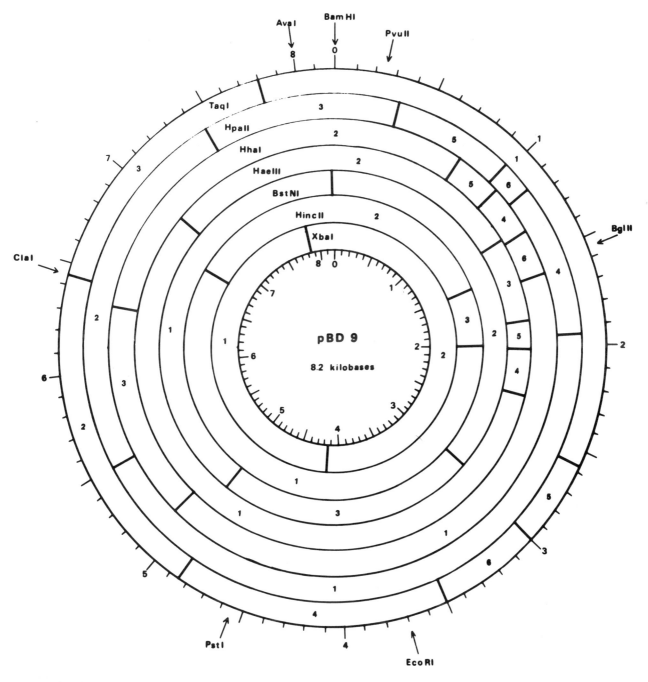

Figure A3.13
Restriction endonuclease cleavage map of pBD9. (From Jalanko and Palva, Gene 14:325, 1981.)

APPENDIX 4

Tables

Table A4.1
DNA polyacrylamide slab-gel reagent volumes

REAGENT		3%			4.5%			7.5%		
	Plate Length (in)	5	6	9	5	6	9	5	6	9
	Spacer Width (mm)	1.5	1.5	1.5	1.5	1.5	1.5	1.5	1.5	1.5
10X Tris-borate (ml)		2.4	3.0	5.0	2.4	3.0	5.0	2.4	3.0	5.0
H_2O (ml)		19.2	24	40.3	18	22.5	37.8	15.6	19.5	33.0
29.2% Acrylamide 0.8% Bis-acrylamide (ml)		2.4	3.0	5.0	3.6	4.5	7.6	6.0	7.5	12.6
10% Ammonium persulfate (ml)		.12	.15	.25	.12	.15	.25	.12	.15	.25
TEMED (μl)		12	15	25	12	15	25	12	15	25
Total (ml)		24	30	50	24	30	50	24	30	50

Table A4.2
Antibiotic resistant transposable elements*

ELEMENT	ASSOCIATED RESISTANCE	SIZE, kb	TERMINAL REPEAT SEQUENCES	REFERENCES
Tn3	Ap^r	4.957	Inverted, 38 bp	Kopecko and Cohen, Proc. Natl. Acad. Sci. USA 72, 1373 (1975) Gill et al., J. Bacteriol. 136, 742 (1978) Heffron et al., Cell 18, 1153 (1979) Gill et al., Nature 282, 797 (1979) Chou et al., Proc. Natl. Acad. Sci. USA 76, 4020 (1979) Chou et al., Nature 282, 801 (1979) Kostriken et al., Proc. Natl. Acad. Sci. USA 78, 4041 (1981)
Tn1, Tn2, Tn401, Tn801, Tn802, Tn901, Tn902, Tn1701, Tn2601, Tn2602, Tn2660	Ap^r	5.	Inverted	Hedges and Jacob, Mol. gen. Genet. 132, 31 (1974) Heffron et al., J. Bacteriol. 122, 250 (1975) Bennet and Richmond, J. Bacteriol. 126, 1 (1976) Benedick et al., J. Bacteriol. 129, 809 (1977) Andreoli et al., Mol. gen. Genet. 160, 1 (1978) Yamada et al., J. Bacteriol. 137, 990 (1979) Chaing and Clowes, J. Bacteriol. 142, 668 (1980) Wang et al., Cell 21, 251 (1980)
TnAβ	Ap^r, Sm^r	14.8	Unknown	Hedges et al., Gene 1, 241 (1977)
Tn4	Ap^r, Sm^r Su^r, Hg^r	20.5	Inverted, 140 bp	Kopecko et al., J. Mol. Biol. 108, 333 (1976)
Tn5	Km^r	5.7	Inverted, IS 50	Berg et al., Proc. Natl. Acad. Sci. USA 72, 3628 (1975) Auerswald and Schaller, Cold Spring Harbor Symp. Quant. Biol. 45, 107 (1981) Rothstein et al., Cold Spring Harbor Symp. Biol. 45, 99 (1981).
Tn6	Km^r	4.2	Direct	Berg et al., Proc. Natl. Acad. Sci. USA 72, 3628 (1975)
Tn2350	Km^r	10.5	Direct	Clerget et al., Mol. gen. Genet. 180, 123 (1980)
Tn903, Tn601	Km^r	3.1	Inverted	Sharp et al., J. Mol. Biol. 75, 235 (1973) Nomura et al., Gene 3, 39 (1978) Grindley and Joyce, Proc. Natl. Acad. Sci. USA 77, 7176 (1980) Grindley and Joyce, Cold Spring Harbor Symp. Quant. Biol. 45, 125 (1981) Oka et al., J. Mol. Biol. 147, 217 (1980)
Tn7, Tn71 Tn72	Tp^r, Sm^r Sp^r	13.5	Unknown	Barth and Datta, J. Gen. Microbiol. 102, 129 (1977) Aymes and Smith, J. Gen. Microbiol. 107, 263 (1978) Datta et al., Cold Spring Harbor Symp. Quant. Biol. 45, 45 (1981)
Tn732, Tn733	Gm^r	10.7 8.7	Unknown	Nugent et al., Nature 282, 422 (1979) Datta et al., Cold Spring Harbor Symp. Quant. Biol. 45, 45 (1981)

ELEMENT	ASSOCIATED RESISTANCE	SIZE, kb	TERMINAL REPEAT SEQUENCES	REFERENCES
Tn1691	Gm^r, Sm^r Su^r, Cm^r, Hg^r	13.5	Inverted, <100 bp	Rubens et al., J. Bacteriol. 140, 713 (1979) Farrar, J., Antimicrobial. Chemother. 6, 4 (1980)
Tn1699, Tn1700	Gm^r, Ap^r Km^r	9.3	Inverted, <100 bp	Rubens et al., J. Bacteriol. 140, 713 (1979) Farrar, J., Antimicrobial Chemother. 6, 4 (1980)
Tn551	Er^r	5.3	Inverted, 35 bp	Kahn and Novick, Plasmid 4, 148 (1980) Novick et al., Cold Spring Harbor Symp. Quant. Biol. 45, 67 (1981) Pattee, J. Bacteriol. 145, 479 (1981)
Tn554	Er^r, Sp^r	6.2	Unknown	Phillips and Novick, Nature 278, 476 (1979) Novick et al., Cold Spring Harbor Symp. Quant. Biol. 45, 67 (1981) Krolewski et al., J. Mol. Biol. 152, 19 (1981)
Tn917	Er^r	5.1	Unknown	Tomich et al., Cold Spring Harbor Symp. Quant. Biol. 43, 1217 (1979) Tomich et al., J. Bacteriol. 141, 1366 (1980)
Tn9, Tn981	Cm^r	2.5–2.7	Direct, IS1	Rosner and Guyer, Mol. gen. Genet. 178, 111 (1980) Reif, Mol. gen. Genet. 177, 667 (1980)
Tn10, TnD, Tn-tet	Tc^r	9.3	Inverted, IS10	Kleckner et al., J. Mol. Biol. 97, 561 (1975) Kleckner et al., Genetics 90, 427 (1978) Kleckner and Ross, Cold Spring Harbor Symp. Quant. Biol. 43, 1233 (1979) Kaulfers et al., J. Gen. Microbiol. 105, 243 (1978) Bennet et al., Plasmid 3, 135 (1980)
Tn804	Tc^r	16	Inverted, <200 bp	Jorgensen et al., Antimicrobiol. Agents Chemother. 18, 200 (1980)
Tn916	Tc^r	15	Unknown	Franke and Clewell, J. Bacteriol. 145, 494 (1980) Franke and Clewell, Cold Spring Harbor Symp. Quant. Biol. 45, 77 (1981)
Tn1721, Tn1771	Tc^r	11.4	Inverted, 38 bp	Choi et al., Cold Spring Harbor Symp. Quant. Biol. 45, 64 (1981) Schmitt et al., Mol. gen. Genet. 172, 53 (1979) Schmitt et al., Cold Spring Harbor Symp. Quant. Biol. 45, 59 (1981) Schoffl et al., Mol. gen. Genet. 181, 87 (1981)
Tn501	Hg^r	8.2	Inverted, 38 bp	Stanisich et al., J. Bacteriol. 129, 1227 (1977) Bennet et al., Mol. gen. Genet. 159, 101 (1978) Brown et al., Nucleic Acids Res. 8, 1933 (1980) Choi et al., Cold Spring Harbor Symp. Quant. Biol. 45, 64 (1981)

Key: Ap = ampicillin; Sm = streptomycin; Su = sulfonamides; Hg = mercury; Km = kanamycin; Tp = trimethoprim; Sp = spectinomycin; Gm = gentamycin; Cm = chloramphenicol; Er = erythromycin; Tc = tetracycline.

*Modified from N. Kleckner (1981). Ann. Rev. Genet. 15, 341.

Table A4.3
Nucleic acid modifying enzymes

ENZYME	ACTIVITIES	APPLICATIONS
1) E. coli DNA Polymerase I	DNA-dependent DNA synthesis, 3'-5'-exonuclease, 5'-3'-exonuclease	Nick translation, synthetic polymer synthesis
2) E. coli DNA Polymerase I Klenow fragment	DNA-dependent DNA synthesis, 3'-5'-exonuclease	Dideoxynucleotide DNA sequencing, Filling in 3' recessed termini, cDNA synthesis
3) T4 DNA Polymerase	DNA dependent DNA synthesis, 3'-5'-exonuclease	Filling in 3' recessed termini, 3' end labeling, ds cDNA synthesis
4) M. luteus DNA Polymerase	Similar to E. coli DNA Polymerase I	Poly(dA-dT)-Poly(dA-dT) synthesis, DNA synthesis on RNA template
5) Terminal deoxynucleotidyl transferase	Addition of dNTP's to 3'-hydroxyl residue of duplex DNA	3'-Homopolymer tailing, synthesis of primers, end-group labeling of 3' termini
6) T4 DNA Ligase	ATP-dependent formation of phosphodiester bond between 5'-phosphate and 3'-hydroxyl residues in duplex DNA	Cohesive ligation, blunt-end ligation
7) E. coli DNA Ligase	DPN-dependent formation of phosphodiester bond between 5'-phosphate and 3'-hydroxyl residues in duplex DNA	Cohesive ligation
8) T4 Polynucleotide Kinase	Transfer of P_i from ATP to the 5' hydroxyl terminus of polynucleotides	^{32}P end labeling of DNA and RNA for nucleotide sequencing
9) AMV Reverse Transcriptase	RNA-dependent DNA synthesis, Degradation of RNA in RNA:DNA hybrid	cDNA synthesis from RNA
10) E. coli RNA Polymerase	DNA-dependent RNA synthesis	*In vitro* RNA synthesis
11) Polynucleotide Phosphorylase	Reversible phosphorolysis/polymerization of polyribonucleotides	Synthesis of polyribonucleotides
12) E. coli poly(A) Polymerase	Polymerization of rA to 3' hydroxyl terminus of RNA	Addition of poly(A) to RNA to allow cDNA synthesis
13) Bacterial or Calf Alkaline Phosphatase	Dephosphorylation of 5'-terminus of RNA and DNA	Dephosphorylation of DNA to prevent ligation, Dephosphorylation prior to end-labeling with polynucleotide kinase

ENZYME	ACTIVITIES	APPLICATIONS
14) S$_1$ Nuclease P$_1$ Nuclease Mung Bean Nuclease	Single-strand specific nuclease	Removal of protruding single strand termini, Hybridization studies, DNA structural probes
15) Exonuclease III	Removal of 3' and 5' mononucleotides from DNA	Production of DNA polymerase substrates, DNA sequencing, DNA mutagenesis
16) Tobacco Acid Pyrophosphatase	ATP hydrolysis	Removal of cap structure from 5' end of mRNA
17) Nuclease BAL-31	Simultaneous exonucleolytic degradation of both 3' and 5' termini of duplex DNA, Single-stranded endodeoxyribonucleolytic activity	Progressively shorten DNA restriction fragments from both ends
18) T4 RNA Ligase	Joining of 3'-hydroxyl terminus of RNA or DNA with a 5'-phosphoryl RNA or DNA molecule	3'-end labeling of RNA and DNA, Construction of defined sequence polymers
19) E. coli RNaseH	Degradation of RNA in RNA:DNA hybrid	Removal of RNA from RNA:DNA hybrid, Removal of poly(A) from mRNA, Site-specific cleavage of RNA

20) Base-specific Ribonucleases
    B. cereus
    PHY M
    T$_1$, U$_1$, N$_1$
    U$_2$
    M$_1$, Phy I
    T$_2$
    CL3
	Site specific RNA cleavage
	Up↓N, Cp↓N
	Up↓N, Ap↓N
	Gp↓N
	Ap↓N
	All *except* adjacent to C residues
	All phosphodiester bonds, preference for A
	Cp↓N, Up↓N
		RNA sequencing

21) E. coli RNase III	Double-stranded RNA degradation	Analysis of RNA processing
22) DNA Topoisomerase	Removes superhelical turns from DNA	Analysis of DNA topology
23) T4 Helix Destabilizing Protein (gene 32 protein)	Binds to single-stranded DNA	Heteroduplex mapping, Detection of low T$_m$ sequences
24) recA protein	ATP dependent reassociation of complementary single stranded DNA	Site specific mutagenesis of DNA

Table A4.4
10X enzyme reaction buffers

STOCK SOLUTIONS	CONC. REAGENT (M)		AluI, BglII, HaeII, III	BamHI	EcoRI	EcoRI*[d]	BglIII	HincII	HindIII	HpaI	PstI	SalI	SmaI	DNA Polymerase (E. coli)	Polynucleotide Kinase (T4)	DNA Polymerase (T4)	DNA Ligase (T4)
	ddH₂O	ml[a]	43.80	38.50	0.0	38.0	0.0	36.80	31.45	17.30	10.0	27.8	32.0	22.00	7.5	4.1	30.0
1.0 TRIS-HCl		mM[b]/ml	60/3.0	200/10.0	1000/50.0	200/10.0	1000/50.0	60/3.0	200/10.0	500/25.0	200/10.0	80/4.0	150/7.5	500/25.0	700/35.0	670/33.5	200/10.0
		pH	8.0	7.0	7.5	8.8	8.0	8.0	7.5	7.5	7.5	7.5	8.0	7.5	7.5	8.8	7.5
1.0 MgCl₂		mM/ml	60/3.0	70[c]/—	50[c]/—	20/1.0	50[c]/—	60/3.0	66/3.3	50/2.5	100/5.0	60/3.0	60/3.0	50/2.5	100/5.0	67/3.35	100/5.0
5.0 NaCl		mM/ml	—	1000[c]/—	1000[c]/—	50/0.5	1000[c]/—	500/5.0	500/5.0	500/5.0	—	1500/15.0	150/7.5	—	—	—	—
1.0 KCl		mM/ml	—	—	—	—	—	—	—	—	500/25	—	—	—	—	166/8.3	—
1.0 (NH₄)₂SO₄		mM/ml	—	—	—	—	—	—	—	—	—	—	—	—	—	—	—
14.0 βME		mM/ml	60/0.22	20/0.07	20/0.07	20/0.07	—	600/2.2	70/0.25	50/0.18	—	—	—	100/0.37	—	100/0.36	—
1.0 DTT		mM/ml	—	—	—	—	—	—	—	—	—	—	—	—	50/2.5	—	100/5.0
0.1 EDTA		mM/ml	—	—	—	—	—	—	—	—	—	0.20/0.1	—	—	—	0.07/0.035	—
5 mg/ml BSA		mg/ml/ml	—	—	—	—	—	—	—	—	—	0.50/0.1	—	0.50/0.1	—	1.7/0.34	—

[a] Figures shown are the number of milliliters of solvent/solute required to give a final volume of 50.0 ml.
[b] Figures are based on a final concentration of 10^{-3} Molar.
[c] Reagent cannot be added by means of a solution. The weight of dry chemical to be added is 0.238 g. of MgCl₂ (50 mM), and 2.92 g. of NaCl (1000 mM).
[d] The EcoRI* cleavage reactive is stimulated by the presence of 20% glycerol.
βME = 2-mercaptoethanol; DTT = Dithiothreitol; BSA = Bovine serum albumin.

Table A4.5
Restriction endonucleases*

ENZYME	SEQUENCE	λ	pBR322	MICROORGANISM
AccI	GT↓(^C_A)(^T_G)AC	7	2	*Acinetobacter calcoaceticus*
AcyI	GPu↓CGPyC	>14	6	*Anabaena cylindrica*
AluI	AG↓CT	>50	16	*Arthrobacter luteus*
AosI	TGC↓GCA	>10	4	*Anabaena oscillaroides*
AosII	GPu↓CGPyC	>14	6	*Anabaena oscillaroides*
ApyI	CC↓(^A_T)GG	>35	6	*Arthrobacter pyridinolis*
AsuI	G↓GNCC	>30	15	*Anabaena subcylindrica*
AsuII	TT↓CGAA	>5	0	*Anabaena subcylindrica*
AvaI	C↓PyCGPuG	8	1	*Anabaena variabilis*
AvaII	G↓G(^A_T)CC	>17	8	*Anabaena variabilis*
BalI	TGG↓CCA	15	1	*Brevibacterium albidum*
BamHI	G↓GATCC	5	1	*Bacillus amyloliquefaciens* H
BamHII	GAA(N)$_2$↓(N)$_3$TTC	>20	>7	*Bacillus amyloliquefaciens*
BamNx	G↓G(^A_T)CC	>17	8	*Bacillus amyloliquefaciens* N
BbeI	GGCGC↓C	?	4	*Bifidobacterium breve*
BclI	T↓GATCA	7	0	*Bacillus caldolyticus*
BglI	GCC(N)$_4$↓NGGC	22	3	*Bacillus globigii*
BglII	A↓GATCT	6	0	*Bacillus globigii*
BluI	C↓TCGAG	1	0	*Brevibacterium luteum*
BpeI	A↓AGCTT	7	1	*Bordetella pertussis*
BstI	G↓GATCC	5	1	*Bacillus stearothermophilus* 1503-4R
BstEII	G↓GTNACC	12	0	*Bacillus stearothermophilus* ET
BstNI	CC↓(^A_T)GG	>35	6	*Bacillus stearothermophilus* N
BstPI	G↓GTNACC	12	0	*Bacillus stearothermophilus*
BsuRI	GG↓CC	>50	22	*Bacillus subtilis* R
BvuI	GPuGCPy↓C	7	2	*Bacillus vulgatis*
CauII	CC↓(^G_C)GG	>30	?	*Chloroflexus aurantiacus*
ClaI	AT↓CGAT	14	1	*Caryophanon latum* L
CltI	GG↓CC	>50	22	*Caryophanon latum*
DdeI	C↓TNAG	>50	8	*Desulfovibrio desulfuricans* Norway strain
DpnI	GA↓TC	(only cleaves methylated DNA)		*Diplococcus pneumoniae*
EcaI	G↓GTNACC	12	0	*Enterobacter cloacae*
EcoRI	G↓AATTC	5	1	*Escherichia coli* RY13
EcoRI'	PuPuA↓TPyPy	>10	15	*Escherichia coli* RY13
EcoRI*	(See Chapter 4)	?	9	*Escherichia coli* RY13

*Restriction endonucleases without known cleavage sites were not included in this list.

(continued)

ENZYME	SEQUENCE	λ	pBR322	MICROORGANISM
EcoRII	↓CC(^A_T)GG	>35	6	Escherichia coli R245
FnuAI	G↓ANTC	>50	10	Fusobacterium nucleatum A
FnuCI	↓GATC	>50	22	Fusobacterium nucleatum C
FnuDI	GG↓CC	>50	22	Fusobacterium nucleatum D
FnuDII	CG↓CG	>50	23	Fusobacterium nucleatum D
FnuDIII	GCG↓C	>50	31	Fusobacterium nucleatum D
FnuEI	↓GATC	>50	22	Fusobacterium nucleatum E
Fnu4HI	G↓CNGC	>50	42	Fusobacterium nucleatum 4H
FspAI	G↓GTNACC	12	0	Flavobacterium species
GdiI	AGG↓CCT	?	1	Gluconobacter dioxyacetonicus
GdiII	Py↓GGCCG	?	?	Gluconobacter dioxyacetonicus
HaeI	(^A_T)GG↓CC(^A_T)	?	7	Haemophilus aegyptius
HaeII	PuGCGC↓Py	>30	11	Haemophilus aegyptius
HaeIII	GG↓CC	>50	22	Haemophilus aegyptius
HapII	C↓CGG	>50	26	Haemophilus aphrophilus
HgaI	GACGC(N)$_5$↓ CTGCG(N)$_{10}$↓	>50	?	Haemophilus gallinarum
HgiAI	G(^A_T)GC(^A_T)↓C	20	8	Herpetosiphon giganteus Hpg5
HgiBI	G↓G(^A_T)CC	>17	8	Herpetosiphon giganteus Hpg5
HgiCI	G↓GPyPuCC	?	9	Herpetosiphon giganteus Hpg9
HgiCII	G↓G(^A_T)CC	>17	8	Herpetosiphon giganteus Hpg9
HgiCIII	G↓TCGAC	2	1	Herpetosiphon giganteus Hpg9
HgiDI	GPu↓CGPyC	>14	6	Herpetosiphon giganteus Hpa2
HgiDII	G↓TCGAC	2	1	Herpetosiphon giganteus Hpa2
HgiEI	G↓G(^A_T)CC	>17	8	Herpetosiphon giganteus Hpg24
HgiGI	GPu↓CGPyC	>14	6	Herpetosiphon giganteus Hpa1
HhaI	GCG↓C	>50	31	Haemophilus haemolyticus
HincII	GTPy↓PuAC	34	2	Haemophilus influenzae R$_C$
HindIII	A↓AGCTT	7	1	Haemophilus influenzae R$_d$
HinfI	G↓ANTC	>50	10	Haemophilus influenzae R$_f$
HpaI	GTT↓AAC	13	0	Haemophilus parainfluenzae
HpaII	C↓CGG	>50	26	Haemophilus parainfluenzae
HphI	GGTGA(N)$_8$↓ CCACT(N)$_7$↑	>50	12	Haemophilus parahaemolyticus
HsuI	A↓AGCTT	7	1	Haemophilus suis
KpnI	GGTAC↓C	2	0	Klebsiella pneumoniae

ENZYME	SEQUENCE	λ	pBR322	MICROORGANISM
MboI	↓GATC	>50	22	Moraxella bovis
MboII	GAAGA(N)₈↓ CTTCT(N)₇↑	>50	?	Moraxella bovis
MlaI	TT↓CGAA	7	0	Mastigoladus laminosus
MnoI	C↓CGG	>50	26	Moraxella nonliquefaciens
PstI	CTGCA↓G	18	1	Providencia stuartii 164
PvuII	CAG↓CTG	15	1	Proteus vulgaris
RsaI	GT↓AC	>50	3	Rhodopseudomonas sphaeroides
RshI	CGAT↓CG	4	1	Rhodopseudomonas sphaeroides
RruI	AGT↓ACT	4	1	Rhodospirillum rubrum
RruII	CC↓(^A_T)GG	>35	6	Rhodospirillum rubrum
SacI	GAGCT↓C	2	0	Streptomyces achromogenes
SacII	CCGC↓GG	3	0	Streptomyces achromogenes
SalI	G↓TCGAC	2	1	Streptomyces albus G
Sau3A	↓GATC	>50	22	Staphylococcus aureus 3A
Sau96I	G↓GNCC	>30	15	Staphylococcus aureus PS96
SfaI	GG↓CC	>50	22	Streptococcus faecalis var. zymogenes
SlaI	C↓TCGAG	1	0	Streptomyces lavendulae
SmaI	CCC↓GGG	3	0	Serratia marcescens S_b
SphI	GCATG↓C	4	1	Streptomyces phaeochromogenes
SstI	GAGCT↓C	2	0	Streptomyces stanford
SstII	CCGC↓GG	3	0	Streptomyces stanford
TaqI	T↓CGA	>50	7	Thermus aquaticus YTI
ThaI	CG↓CG	>50	23	Thermoplasma acidophilum
TthHB8	T↓CGA	>50	7	Thermus thermophilus HB8
Tth111I	TGACN↓NNGTC	2	2	Thermus thermophilus 111
Tth111II	CAAPuCA(N)₁₁↓ GTTPyGT(N)₉↑	>25	5	Thermus thermophilus 111
XbaI	T↓CTAGA	1	0	Xanthomonas badrii
XhoI	C↓TCGAG	1	0	Xanthomonas holcicola
XhoII	Pu↓GATCPy	>20	8	Xanthomonas holcicola
XmaI	C↓CCGGG	3	0	Xanthomonas malvacearum
XmaIII	C↓GGCCG	2	1	Xanthomonas malvacearum
XpaI	C↓TCGAG	1	0	Xanthomonas papavericola

Table A4.6
Supplies and equipment

The following items are frequently used in molecular cloning work.

ITEM	CATALOG #	SOURCE
EQUIPMENT:		
1.5 ml conical microfuge tubes	72690	Sarstedt, Inc. 745 Alexander Rd. Princeton, NJ 08540
		BioRad Laboratories PO Box 708 Rockville Center, NY 11571
Pipetman, P20, P200 (Gilson)	—	West Coast Scientific PO Box 2947 Rockridge Station Oakland, CA 94618
		Rainin Instrument Corp. Inc. Mack Road Woburn, MA 01801
Pipetman tips	OPS1035	Outpatient Services, Inc. 1320 Scott St. Petaluma, CA 94952
	1137BK	Evergreen Scientific Co. 2300 E. 49th St. Los Angeles, CA 90058
	—	BioRad Laboratories PO Box 708 Rockville Center, NY 11571
Eppendorf test tube racks (polypropylene collection plates)	26107	Gilson Medical Electronic PO Box 27 Middleton, WI 53562
Eppendorf centrifuge	C3514-1	American Scientific Products, Inc. 1430 Waukegan Rd. McGaw Park, IL 60085 or
Temp-Block Module	H2025-1	American Scientific Products, Inc. 100 Raritan Center Parkway Eddison, NJ 08818
Electrophoresis Power Supply Models 493 or 494	—	Instrumentation Specialities Co. PO Box 5347 Lincoln, NE 68505
Micro Fractionator Gilson, Model RC-80K	—	Gilson Medical Electronics, Inc. PO Box 27 Middleton, WI 53562

ITEM	CATALOG #	SOURCE
Chromatography columns	—	Kontes Glass Co. Vineland, NH 08360
		BioRad Laboratories 2200 Wright Avenue Richmond, CA 94804 or BioRad Laboratories PO Box 708 Rockville Center, NY 11571
Biogel A-50 M 100–200 mesh (gel filtration)	151-1340	BioRad Laboratories 2200 Wright Avenue Richmond, CA 94804 or
Dowex AG 50W-X8 100–200 mesh (cation exchange resin)	142-1441	BioRad Laboratories PO Box 708 Rockville Center, NY 11571
Acrylamide Bis-Acrylamide and Temed	161-0101 161-0200 161-0800	
Agarose	162-0100	
ANTIBIOTICS:	Ampicillin	Bristol Laboratories, Inc. Div. of Bristol-Myers Company Syracuse, NY 13201
	Chloramphenicol and Tetracycline	Sigma Chemical Co. PO Box 14508 St. Louis, MS 63178
RESTRICTION ENZYMES:	Bethesda Research Laboratories, Inc. 411 North Stonestreet Avenue Rockville, MD 20850 (800) 638-8992	
	Boehringer Mannheim 7941 Castleway Drive PO Box 50816 Indianapolis, IN 46250 (800) 428-5433	
	New England Biolabs, Inc. 283 Cabot St. Beverly, MA 01915 (617) 927-5054	
	P-L Biochemicals, Inc. 1037 West McKinley Avenue Milwaukee, WI 53205 (800) 558-7110	
	Biochemical Product Division Worthington Diagnostic Systems Inc. A Flow General Company Freehold, NJ 07728	

Index

A50 column, 41
ABM paper, 110
Agarose gel electrophoresis. *See* Electrophoresis
Agrobacterium tumefaciens, 159, 160, 183
Alkaline phosphatase, 21, 86
Amplification of plasmid DNA, 5
Antibiotic resistance, 27
Antibody, 76, 109, 110
A-protein, 110
Autoradiography:
    detection of clones, 109
    Southern blot hybridization, 74

Bacillus subtilis, 162, 164, 184
Bacterial growth:
    crossfeeding, 104
    feeder colonies, 28, 29
    microcolonies, 28
    minimum inhibitory concentration (MIC), 28
    slants, 33
    stabs, 33
    standard growth curve, 30, 35
    storage of strains, 33
Bacteriophage:
    λ, 12
    λ Charon 4A, 13, 15
    λ Charon 30, 15
    λgtWES·λB, 13, 15
    M13, 9
    T4, 81
    T7, 57

Bacteriophage vector. *See* Vectors
Blot-transfer, 74, 109
Blunt ends, 57

Cesium chloride density gradient, 37, 155
Charon phages, 13–15
(Chi) χ1776 strain of *E. coli*, 94
Chloramphenicol:
    amplification of plasmid DNA, 5, 156
    resistance gene, 8, 136
Chloramphenicol acetyl transferase (CAT), 136, 187
Chromosomal DNA, isolation of, 38, 45, 162, 167, 172
Chromosomal transfer, 92
Cleared lysate, 155
Cloning strategies, 1, 19
Cloning vectors. *See* Vectors
CNBr activate paper, 110
Cohesive ends on DNA fragments, 57
Colony hybridization, 109
α-complementation, 9, 10
Complementation of phenotype, 21
Concatemers, 16, 71, 83, 85
Conjugation, 91
Containment:
    biological, 15
    physical, xviii, 15
Cos site, 12, 16
Cosmid, 16

INDEX

Dialysis, 200
Dihydrofolate reductase, 103
Dilution factor (df), 36
DNA:
   determination of concentration, 42, 47
   dideoxynucleotide sequencing, 11, 12
   forms I, II, III, 70
   isolation of, 38–50
   regulatory sequences, 129
   supercoiled, 37, 39, 70
DNA isolations:
   B. subtilis chromosomal, 162
   B. subtilis plasmid, 164
   CaCl-PdI density gradient, 155
   E. coli chromosomal, 45
   Drosophila, 172
   electroelution from gels, 178
   high molecular weight plasmids, 159
   large plasmid miniscreen, 160
   low-melt agarose gels, 173
   phage lambda, 165
   plasmid miniscreen, 50
   preparative agarose gel, 175
   yeast, high molecular weight, 167
   yeast, plasmid, 170, 171
DNA kinase, 57
DNA ligase. *See* Ligase
DNA modifying enzymes (table), 224
DNA polymerase, 60
DNA sequencing, 11, 12
DNase I, 165
Dowex AG50W-X8, 40, 153, 154, 158, 174
Drosophila, 172
Dyad symmetry, 55, 133, 134

E. coli strains:
   AB257, 32
   BE42, 32
   BHB2688, 181
   BHB2690, 181
   χ1776, 94, 179
   CSR603, 192
   JM101, 10, 180
   JM103, 10
   LA6, 32
   P678-54, 194
   RR1, 32
   RR1 (pBR329), 32
   RR1 (pPV33), 32
   RR1 (pPV501), 32
*Eco*RI* activity, 58
Efficiency of plating (EOP_{50}), 135
Electrophoresis:
   agarose, 68
   artifacts, 72
   gel systems, 71
   and hybridization analysis, 74
   polyacrylamide, 68
   principles, 67
   tracking dyes, 64
   visualizing DNA, 69
Endonuclease. *See* Restriction endonuclease
Enhancer, 130
Ethidium bromide:
   intercalation, 39
   staining of DNA, 69
   use in DNA purification, 37
Expression, gene:
   immunochemical detection of, 109
   vectors, 22

F, fertility factor, 91
Flushends. *See* Blunt ends
Fused translation products, 11, 22, 23, 138

β-galactosidase, 9
Gel staining, 78
Gene bank, 19
Gene cloning, 1
Genetic engineering, 1
Globin, human genes, 19, 20
Grunstein and Hogness procedure, 109

Hfr, 92
Homopolymer tailing, 22, 23, 57
Host controlled restriction/modification, 53, 96
Hybridization, 22, 74, 109

i, 83
Immunochemical detection of clones, 109
in vitro packaging, 16, 95, 181
Insertional activation, 134
Insertional inactivation, 4, 5, 113
Intercalating dyes, 39, 40
Internal promoters, 130
Intervening sequences, 20
Introns, 20
IPTG, 10, 180
Irradiation, UV. *See* UV irradiation
Isoschizomers, 5

j, 83
j/i ratio, 84

Klenow fragment of DNA polymerase I, 60
Klett-Summerson colorimeter 36

Lac promoter, 10, 135, 136
β-lactamase, 22, 29
Ligase, DNA, 1, 81
  cloned gene, 82
  cofactor requirements, 81
  mechanism, 82
Ligation of DNA:
  blunt end, 86
  cohesive end, 82
  factors affecting, 83
  j/i ratio, 84
  use of alkaline phosphatase, 86
Linkers, synthetic, 60
Lysate, cleared, 155

Map, restriction:
  lambda, 217
  M13 mp73, 10
  M13 mp7, 214
  M13 mp7, mp8, mp9, mp10, mp11, 215
  pBD9, 219
  pBR322, 204
  pBR327, 207
  pBR328, 210
  pBR329, 211
  pUB110, 218
  pUC7, pUC8, pUC9, 216
Maxicells, 117, 192
Media:
  freezing medium, 151
  Hershey, 192
  K medium, 192
  LB, 149
  LB + DAPT, 179
  M9, 150
  M523, 161, 184
  SPI, SPII, 185
  yeast complete, 153
  yeast minimal, 151, 153
  YEB, 183
  YT, 149
Methylation of DNA, 53, 56
Minicells, 115, 194
Modification, host specified, 53
Molecular cloning, 1

Nin deletion, 13
Nitrocellulose filters, 74, 109, 132
Northern blot, 76
Nucleic acid modifying enzymes (table), 224

Origin of replication, 5

Packaging, *in vitro*, 16, 95, 181
Phage. *See* Bacteriophage

Phenotypic characterization, 33, 103, 119
Phosphatase, alkaline, 21, 86
Plaque hybridization, 109
Plasmids:
  amplification, 5, 156
  as cloning vectors, 3
  enrichment, 39
  phenotypic traits, 4
  purification. *See* DNA isolations
  relaxed, 4
  stringent, 4
Plasmids, specific:
  ColE1, 4
  pBR313, 5, 6
  pBR322, 4, 6, 204, 205
  pBR324, 7, 8
  pBR325, 7, 8
  pBR327, 7, 8, 207, 208
  pBR328, 7, 8, 210
  pBR329, 7, 8, 211, 212
  pI47, 22
  pLP1, 136
  pMB8, 6
  pMB9, 6
  pPV33, 135
  pPV501, 137
  pSa, 57
  pSC101, 4-6
  pTR262, 20
  pUB110, 218
  pUC, 9, 12
  RSF2124, 5, 6
Plasmid vectors. *See* Vectors
Polyacrylamide gel electrophoresis. *See also* Electrophoresis
  DNA gels, 221
  protein gels, 197
Polymerase, DNA, 60
Polymerase, RNA, 130
Polyvinyl, disk, 110
Positive selection:
  complementation, 21
  gene activation, 20, 134
  λ packaging, 14
Preproinsulin, 22, 23
Pribnow box, 130, 131
Probe, 22, 75
Promoter, 130
  lac promoter, 10, 135, 136
  Tc^r promoter, 8
Promoter-probe vectors, 18, 134
Propidium iodide, 39, 155

Reading frame, 137
Replica plating, 105
Restriction/modification, host controlled, 53

INDEX

Restriction endonuclease, 1
  cofactor requirements, 55
  digestion conditions and specificity, 58
  frequency of occurrence, 56
  generating cleavage sites, 59
  nomenclature, 55
  purification of, 58
  table of, 227
  types, 55
Restriction enzyme buffers (table), 226
Restriction mapping, 106
Reverse transcriptase, 20
RNA modifying enzymes (table), 224
RNA polymerase, 130
  binding assay, 131

Saccharomyces cerevisiae. See Yeast
Screening methods, 22
Segregation, 27
Selection vs. screen, 20
Sequencing of DNA, 11, 12
Sequence, nucleotide:
  pBR322, 205
  pBR327, 208
  pBR329, 212
Single colony isolate, 33, 34
Sonication, 188
Southern blot, 74, 76. *See also* Hybridization
Specialized cloning vectors. *See* Vectors
Spheroplasts, 94, 96
Stabs, agar, 33, 149
Sticky ends. *See* Cohesive ends on DNA fragments
Subcloning, 111

Terminal transferase, 22, 57
Terminator-probe vectors, 18, 136
Terminators, transcriptional, 133
T_m, 83, 87

Tra genes, DNA transfer, 91
Transcription:
  initiation, 130
  termination, 133
Transduction, 93
Transfection. *See* Transformation
  λ, 181
  M13, 180
Transformation:
  A. tumefaciens, 183
  bacterial, 93
  B. subtilis, 94, 184
  E. coli, 99
  E. coli χ1776, 94, 179
  effect of DNA structure, 96
  efficiency, 96
  transfection, λ, 181
  transfection, M13, 180
  to verify phenotype, 103, 106
  yeast, 96, 186
Transposon, transposable element, 4, 114
Trimethoprim resistance, 103

UV irradiation, maxicells, 117, 193
  DNA gels, 69, 78

Vectors:
  cosmids, 16
  criteria for design, 4
  double-stranded bacteriophage, 12
  plasmids, 4
  replacement, 14
  single-stranded bacteriophage, 9
  table of, 17
Volume correction, 36

Western blot, 76

X-gal, 10

Yeast, 96